LEADING AI WELL

An Executive's Guide to AI Strategy

Jason D. Baker

In Collaboration with Claude

WATERTREE PRESS

LEADING AI WELL:
An Executive's Guide to AI Strategy

Copyright © 2026 by Jason D. Baker

Published by Watertree Press, LLC
PO Box 16763, Chesapeake, VA 23328
http://www.watertreepress.com

Publisher's Cataloging-in-Publication Data

Names: Baker, Jason D., author.
Title: Leading AI Well: An Executive's Guide to AI Strategy / Jason D. Baker.
Description: Chesapeake, VA: Watertree Press LLC, 2026
Identifiers: ISBN: 978-0-9911046-2-8 (pbk.) | LCCN: 2026937602
Subjects: LCSH: Artificial intelligence. |Management—Technological innovations. |Business enterprises—Information technology—Management.
BISAC: BUSINESS & ECONOMICS / Management. | BUSINESS & ECONOMICS / Decision-Making & Problem Solving. | COMPUTERS / Artificial Intelligence / General.
Classification: LCC HD57.7.B25 | DDC 658.4–dc23

Library of Congress Control Number: 2026937602

CONTENTS

Introduction

THE WAYMO MOMENT

The first time I climbed into a Waymo, I felt like I was stepping into science fiction. No driver; just an empty seat where a human should have been. And a cheerful welcome screen addressing me by name.

I watched the steering wheel turn on its own as we pulled into traffic. I tracked the display showing what the car "saw"—the route ahead in green, other vehicles in blue, pedestrians glowing. When a delivery truck double-parked ahead, the car signaled, changed lanes, and continued as if mildly inconvenienced rather than genuinely challenged. I pulled out my phone and excitedly sent a "hey, look at this!" video to family and friends.

By the third ride, it felt routine. I checked my messages and barely glanced at the road. The revolution had become Tuesday.

That shift—from astonished to average in the space of a few days— is the real story of where we are with AI. The extraordinary is becoming ordinary so quickly that we're failing to register how much has already changed.

TRY THIS

Before reading further, write down three things you believe AI cannot do well. Be specific. "AI can't understand our customers' emotional needs." "AI can't make judgment calls about brand voice." "AI can't handle the nuance in our regulatory environment." Keep this list somewhere you can find it. We'll revisit it at the end.

The Revolution Isn't Coming. It's Here.

Most executives I talk to intellectually acknowledge AI's importance. They've read the articles, attended the conferences, and added "AI strategy" to their board agendas. Yet when I ask about specific capabilities—what can AI really do right now?—their answers reveal a significant gap between perception and reality.

They think AI means chatbots. But that's a bit like thinking "food" means "pizza"—true as far as it goes but missing most of the picture.

While executives debate whether AI will eventually matter, autonomous vehicles are navigating rush-hour traffic in San Francisco. Dark factories in China produce goods around the clock with no human workers on the floor. AI systems design novel proteins, optimize global supply chains, execute complex trades, and fly military drones. Code that would have taken a team of engineers weeks to write now emerges in minutes. Contracts that once required hours of review are analyzed in seconds.

This isn't a preview of coming attractions. It's what's playing right now, today, in industries all around us.

The question isn't whether AI will transform our industries. It's whether we'll be the ones leading the transformation or the ones being transformed.

Yes, And

In improvisational theater, there's a foundational rule called "yes, and." When your scene partner makes an offer—"Look, a dragon!"—you accept it and build on it. "Yes, and it seems to be wearing a tiny hat!" The scene moves forward. Both performers contribute. Something unexpected emerges.

The alternative is blocking. "That's not a dragon, that's a cloud." The scene dies. Your partner is left stranded. Nothing gets built.

I've borrowed "yes, and" as the foundational mindset for this book because it captures exactly what separates leaders who will thrive in the AI era from those who will be left behind.

When someone asks, "Can AI do X?"—write marketing copy, analyze financial statements, generate code, design products—the strategic answer is almost always: "Yes, and here's what that means. And here's what we do about it."

Blocking is tempting. "AI can't really understand context." "That's years away." "Not in our industry." "Our customers prefer the human touch." Each of these statements might even be partially true. But leaders who lead with objections—who block rather than build—will find themselves edited out of the scene entirely.

This doesn't mean naive optimism. "Yes, and" isn't about pretending AI has no limitations or that every application will succeed equally. It's about strategic acceptance. We can't outmaneuver a reality we refuse to acknowledge. We can't build on a foundation we won't stand on.

Throughout this book, I'll invite you to practice "yes, and" thinking—to accept what AI has established and build on it, even when the implications are uncomfortable, even when they threaten assumptions you've held for years.

Who This Book Is For

I wrote this for C-Suite executives and entrepreneurs who recognize AI's strategic importance but need frameworks for thinking clearly and acting decisively without becoming technical experts. If you've used ChatGPT or Claude and thought "this is interesting, but now what?"—this book is for you.

I also wrote this for the senior directors and aspiring executives who will be leading AI initiatives or stepping into C-suite roles. The strategic literacy you build now will define your career trajectory.

This is the AI education your MBA didn't include. When most business schools designed their curricula, AI was a research

3

curiosity, not an operational reality. That gap isn't your fault—but closing it is your responsibility.

ASK AI

Open your preferred AI assistant and ask: "What's the most surprising thing AI systems are doing right now that most people don't know about?"

How This Book Was Written

I should tell you that this book is itself an example of the kind of AI leadership I'm advocating.

The work you're reading is a collaboration between me and Anthropic's Claude. Claude has served as an editorial partner throughout the process—comparing insights to other business books, helping polish frameworks, pressure-testing arguments, drafting and revising prose, and challenging me when my thinking was fuzzy or my examples weren't landing. The ideas are mine. The judgment calls are mine. The responsibility for what's on these pages is mine. But the execution has been a genuine human-AI collaboration.

I'm not hiding this because I'm not ashamed of it. In fact, I think pretending otherwise would undermine the entire premise of the book. If I'm going to ask you to embrace "yes, and" thinking—to accept AI as a collaborator and build on what it offers—I should be willing to model that myself.

This collaboration has made the book better than I could have written alone. Not because Claude knows more than I do about AI strategy—that's not how this works—but because having a tireless thinking partner who can draft, critique, restructure, and iterate at speed changes what's possible. The book you're holding exists in its current form because I practiced what I'm preaching.

Principles Over Procedures

This is a short book by design. You can read it on a cross-country flight or in a focused afternoon. It's not a machine learning textbook, rather it's a guide to essential frameworks, concepts, and action steps that will change how you lead.

You might expect a book on AI leadership to include deep-dives into prompting techniques, agent configuration, or tool comparisons. This one doesn't—deliberately.

The tactical landscape shifts too fast. Specific prompt patterns that work brilliantly today may be obsolete in six months. Tool recommendations date before the ink dries. Detailed configuration guides become historical artifacts (which I've experienced, having authored an early internet book in 1994).

Instead, this book focuses on the strategic layer—the mental models, frameworks, and judgment that remain durable even as tools evolve. We'll cover principles of AI fluency and collaboration. The callouts will give you a taste of effective AI engagement, framed as examples rather than comprehensive tutorials. The goal is to build your judgment about what to do, not create dependency on instructions for how.

The book unfolds in three parts. In Part I, we'll work on seeing clearly—understanding what AI actually is, what it can and can't do, and building your capabilities. In Part II, we'll turn to strategic thinking—how AI reshapes competition, business models, and the nature of work. In Part III, we'll focus on leading responsibly—governance, investment, and execution.

Every section answers three questions: What? So What? Now What? In other words: What is this? Why does it matter? What do you do about it? You'll also find "Try This" and "Ask AI" callouts—brief, actionable prompts inviting you to apply ideas immediately, often using AI itself.

What Will You Do?

By my third Waymo ride, the magic had become mundane. I suspect something similar has already happened with AI in your own experience. You've used it enough that the novelty has worn off—but perhaps not enough to grasp the full scope of what's now possible.

Here's what I want you to hold as we move forward: You're not preparing for an AI future. You're catching up to an AI present. The autonomous vehicle isn't coming—it's already giving rides. The AI assistant isn't a prototype—it's already doing the work. The transformation isn't theoretical—it's already reshaping your industry, whether you see it yet or not.

The only question is whether you'll block or build.

Let's build together. And let's start with the most fundamental shift AI represents—the transition from a world of scarcity to a world of abundance.

Part I

SEEING CLEARLY

Before you can act wisely, you need to see accurately—and you can't see AI from the sidelines.

1

FROM SCARCITY TO ABUNDANCE

What if the thing we've been rationing is suddenly everywhere?

Consider how much strategy depends on scarcity. We hire carefully because talent is limited. We guard data because insights are hard-won. We ration access to expertise because there's only so much to go around. We build competitive moats around proprietary analysis because generating it requires expensive, specialized humans working long hours.

These constraints feel like facts of nature—the way business works. But they're not natural laws. They're artifacts of a particular moment in time. And that moment is ending.

What happens when the constraint disappears? What happens when the thing we've been rationing becomes as available as water from a tap?

This isn't a hypothetical question. It's the question every leader has faced at major technological inflection points throughout history. And most have answered it poorly—not because they were stupid, but because abundance is genuinely hard to comprehend when we've built our world around scarcity.

The Recurring Pattern

The most transformational technologies in human history share a common structure: they take something scarce and make it abundant. This shift—from scarcity to abundance—isn't just one dynamic among many. It's the defining characteristic of technological revolution.

The agricultural revolution made food abundant. For most of human existence, securing enough calories to survive was the central challenge of daily life. Hunting, foraging, and early farming consumed most of humanity's time and attention. Then came systematic agriculture: irrigation, crop rotation, selective breeding, and eventually mechanization. Today, the developed world's food challenge isn't scarcity but overabundance. We have food deserts in countries that throw away nearly one-third of what they produce. The constraint that shaped human civilization for millennia simply...dissolved.

The industrial revolution made physical goods abundant. Before mechanized manufacturing, most objects were handcrafted by skilled artisans. Clothing, furniture, tools—made by human hands, one piece at a time. A single shirt represented hours of spinning, weaving, and sewing. Then came factories, interchangeable parts, and assembly lines. Today, we have fast fashion, disposable electronics, and landfills overflowing with things that cost almost nothing to produce. The scarcity of physical objects that defined pre-industrial life became the abundance that defines our disposable culture. Try giving your grandmother's china to your children. They don't want it. When anything can be replaced, nothing needs to be kept.

The information revolution made knowledge abundant—and it did so in waves. First, writing made information storable and transmissible beyond individual human memory. But for millennia, written documents remained scarce: each copy required a scribe working by hand, and literacy was rare. Then the printing press made text reproducible at scale, democratizing access to knowledge and enabling the Scientific Revolution, the Reformation, and the Enlightenment. Finally, digital technology made information not just reproducible but instantly accessible from anywhere on earth. The sum of human knowledge now fits in your pocket.

Each transition followed a similar arc. First, a new technology emerged. Then, early adopters exploited it while most people failed to grasp its implications. Eventually, the abundance became so pervasive that it reshaped everything: business models, social structures, power dynamics, the nature of work itself. Those who understood the shift early built empires. Those who clung to scarcity assumptions watched their advantages evaporate.

We are now in the midst of another such transition. AI is making intelligence abundant.

What AI Makes Abundant

When I say AI makes "intelligence" abundant, I need to be more specific. Intelligence is a broad term, and AI doesn't replicate all of it equally. What AI makes abundant—right now, today—includes analysis, pattern recognition, content generation, and increasingly, certain forms of reasoning. Let's consider each in turn.

Analysis used to be expensive. Understanding your market, your customers, your operations—all of it required human analysts poring over data, building models, writing reports. A comprehensive competitive analysis might take a team weeks. A detailed financial model required specialists. Now? AI can analyze thousands of documents in minutes. It can process customer feedback at scale, identify patterns in operational data, and generate insights that would have taken human analysts months to produce. The bottleneck has shifted from "can we analyze this?" to "do we know what questions to ask?"

Pattern recognition has undergone a similar transformation. Humans are remarkably good at recognizing patterns, but we're slow, inconsistent, and limited in the complexity we can process. AI systems can now identify patterns in medical imaging that experienced radiologists miss. They can detect fraud by recognizing subtle anomalies across millions of transactions. They can predict equipment failures by finding patterns in sensor data that no human could perceive. In logistics, AI systems optimize delivery routes by recognizing patterns in traffic, weather, and demand that would overwhelm human planners. The pattern recognition that once required rare expertise is becoming a commodity.

Content generation is perhaps the most visible transformation. Writing, designing, coding—activities that once required skilled

11

humans working for hours—can now be accomplished in seconds. Marketing copy, technical documentation, software code, business presentations: AI generates all of these at speeds and scales that would have seemed absurd five years ago. A task that might have occupied a junior employee for a week can now be drafted in minutes.

And reasoning—the domain we once considered uniquely human—is increasingly within AI's reach. Not all reasoning, of course, and not perfectly. But AI systems can now work through complex arguments, evaluate options against criteria, identify logical flaws, and generate strategic alternatives. They can engage in the kind of structured thinking that MBAs are trained to do: analyzing business cases, evaluating trade-offs, constructing recommendations. This isn't replacing human judgment—not yet, not entirely—but it's augmenting and accelerating it in ways that fundamentally change what's possible.

These categories aren't exhaustive and the boundaries keep expanding. But the core point remains: capabilities that were once scarce—expensive to acquire, slow to deploy, limited in scale—are becoming abundant.

TRY THIS

List three capabilities your organization currently treats as scarce—things you ration, prioritize carefully, or can't do as much of as you'd like. For each one, ask: What would change if AI made this abundant next year? What about next week?

Technology Shifts Time

Before shifting to scarcity, there's a caveat worth noting in how technology affects work: it doesn't so much save time as shifts time.

Consider presentations. You used to be able to bring handwritten notes. Now you need polished slides with professional graphics. Yes, it got faster and cheaper to produce professional presentations—but expectations rose to match. The time didn't disappear; it moved from production to refinement.

AI will follow the same pattern. Tasks that once took hours will take minutes—and then we'll raise our standards for what those minutes should produce. The value of AI isn't leisure; it's capability. Once early adopter arbitrage passes, it's more likely to raise the bar for what a week's work looks like than give us longer weekends.

What Remains Scarce

Here's where the strategic thinking gets interesting. When something becomes abundant, value doesn't disappear—it migrates. It moves to whatever remains scarce.

When food became abundant, value migrated to cuisine, experience, and status. When physical goods became abundant, value migrated to design, brand, and craftsmanship. When information became abundant, value migrated to curation and interpretation.

The same migration is happening now. Understanding these dynamics is essential to build sustainable advantage.

Right up front, two fundamental scarcities deserve attention. These aren't skills or attributes—they're constraints that shape everything else.

Time

AI can produce thousands of documents in the time it takes you to read one. That's the abundance. But you still have only twenty-four hours in a day. That's the scarcity that doesn't budge.

Time scarcity means that even in a world of infinite content and instant analysis, attention becomes the binding constraint. The bottleneck shifts from "can we produce this?" to "can anyone absorb it?" From "do we have the analysis?" to "do we have time to act on it?"

For leaders, time scarcity means your judgment about where to direct attention—yours and your organization's—matters more than ever. AI can do more things; you still can't focus on more things.

Trust

Trust is slow to build and fast to destroy. AI doesn't change that equation—if anything, it intensifies it.

In a world where anyone can generate convincing content, authentic trust becomes rarer and more valuable. When deepfakes and synthetic media erode confidence in what we see and hear, established relationships become anchors of credibility.

Trust also carries a tax. Systems and relationships that lack trust cost more—they require more oversight, more verification, more redundancy. Every moment spent double-checking is a moment not spent creating value. Organizations with high internal trust move faster than those that don't. Leaders who've built trust get the benefit of the doubt; those who haven't face friction at every turn.

Trust built on your relationships, reputation, and track record will appreciate in value as AI saturates the market.

As AI makes certain forms of intelligence abundant, value is migrating to what remains genuinely scarce.

Judgment remains scarce. AI can analyze options and generate recommendations, but deciding what matters—which values to prioritize, which risks to accept, which trade-offs to make—requires human judgment. An AI can tell you the expected returns of different strategic choices, but it can't dream of a better future. It can model the consequences of a decision, but it can't own that decision or take responsibility for it.

Relationships remain scarce. AI can simulate conversation, but it cannot form genuine relationships. It cannot build the networks of mutual obligation, shared history, and personal loyalty that still drive much of business. Human connections don't scale, can't be automated, and can't be faked for long—which is precisely what makes them valuable.

Attention remains scarce. In an age of infinite content, human attention becomes the limiting factor. There are still only twenty-four hours in a day, and the human capacity to focus remains stubbornly finite. As AI floods every channel with more content, more analysis, more options, the ability to command genuine human attention becomes more precious. Cutting through the noise isn't getting easier; it's getting harder.

Wisdom, distinguished from intelligence, remains scarce. Intelligence can analyze; wisdom knows which analysis to trust. Intelligence can optimize; wisdom knows what shouldn't be optimized. Intelligence can solve problems; wisdom knows which problems are worth solving. Wisdom integrates knowledge with experience, judgment with humility, capability with ethics. It's the meta-skill that guides how all other skills are deployed. AI can make us more intelligent; it cannot make us wise.

The strategic question for every leader is this: Are you building your competitive position around things that are becoming abundant, or around things that remain scarce?

The Value Proposition Question

This brings us to the central strategic question of the AI era: If AI can do this, why do customers need us?

It's a brutal question, and most executives instinctively resist it. Surely our expertise, our relationships, our brand, our years of experience—surely these must count for something. And they do. But the question isn't whether your current value proposition has merit. The question is whether it will survive contact with abundance.

Consider the consulting industry. For decades, top consulting firms have commanded premium fees for analysis, frameworks, and recommendations. Their value proposition rested on access to specialized expertise, proprietary methodologies, and the concentrated brainpower of very smart people working very hard. But what happens when AI can generate sophisticated analysis in minutes, using frameworks accessible to anyone, when the brainpower is no longer scarce?

15

The same logic applies to every industry. Never assume your value proposition is immune just because it feels essential today. In 1998, it was easy to dismiss Amazon's threat to bookstores. "People like browsing." "They want to hold the book before they buy it." "The bookstore experience can't be replicated online." Every one of these statements was true, and every one of them was irrelevant. Amazon didn't have to replicate the bookstore experience; it had to offer something better on the dimensions that mattered to most customers: selection, convenience, and price. And that's before Amazon entered the market for clothes, shoes, movies, cloud computing, medicine, and more.

The scarcity you've built your business around may be the next thing to become abundant. The question isn't whether this is comfortable to contemplate. The question is whether you'd rather confront it now, when you still have time to adapt, or later, when you don't.

ASK AI

"What scarcity assumptions might be embedded in my company's strategy? Where are we assuming that something will remain scarce that AI might make abundant?" Then ask: "If AI can do what we do, why would customers still need us?" The answers may be uncomfortable. That's the point.

The Dark Side of Abundance

Abundance isn't purely positive. Every abundance transition has brought new problems along with new possibilities, and the AI transition is no exception.

When content creation costs approach zero, garbage gets created and distributed at scale. AI slop. The flood of low-quality, AI-generated content that technically "works" but adds no real value. You've already seen it. The SEO article that reads smoothly but says nothing. The marketing email that's grammatically perfect and completely generic. The report that's professionally formatted and utterly empty.

Just because you can produce infinite content doesn't mean you should. Organizations can become slop factories without realizing it—every channel filled, every request answered, every gap stuffed with content. From a productivity standpoint, this looks like success. From a value standpoint, it's often worse than nothing. AI slop doesn't just fail to help; it actively damages trust, attention, and brand.

The slop problem is also a governance problem, but for now, the point is simply this: abundance requires new disciplines, not just new capabilities. The organizations that thrive won't be those that produce the most AI-generated content; they'll be those that produce the most valuable content, by whatever means. That's a judgment call—one of those scarce things AI can't provide.

A New Strategic Lens

The scarcity-to-abundance framework isn't just an interesting historical pattern. It's a strategic lens that should inform every AI decision you make.

When evaluating an AI initiative, ask: Is this exploiting something AI makes abundant, or protecting something that remains scarce? When assessing competitive threats, ask: Is our position built on scarcity that AI might dissolve? When developing talent, ask: Are we building skills that will become commodities, or skills that will become more valuable as AI proliferates?

The strategies that worked in scarcity will fail in abundance. The advantages built on scarcity will erode.

2

What AI Is (And Isn't)

You've used ChatGPT or Claude. You've seen the demos. You've experimented enough to have opinions about what works and what doesn't. You're not ignorant about AI.

But here's the thing about familiarity: it can actually make it harder to see the whole picture. It's a bit like knowing a city from a single neighborhood. You've walked the streets, found your favorite coffee shop, figured out the subway stop. You're not a tourist anymore. But that comfort with one area can trick you into thinking you understand the whole city, when really you just know your corner of it.

For most executives, that familiar neighborhood is conversational AI—the chatbots and assistants you've typed questions into. And that neighborhood is real and important. But AI is a much larger city than that one district, and it operates by rules that might be quite different from the software you've worked with your entire career.

Think of this as a quick tour. By the end, you'll understand AI well enough to ask good questions, spot overblown claims, and make sound decisions about where and how to deploy it.

Let's start with the lay of the land.

Five Districts of AI

When most people hear "AI," they picture a chatbot. Useful, yes. Representative of the whole category, no.

AI spans multiple territories, each with different capabilities, use cases, and implications for your business. Understanding these differences will help you see opportunities and threats that remain invisible to executives who think AI begins and ends with text conversations. For this tour, we'll consider five different AI districts.

Conversational AI is the neighborhood you know best. These are the chatbots and virtual assistants—ChatGPT, Claude, Gemini, Copilot—that you interact with through natural language. You type or speak; they respond. They can answer questions, draft documents, brainstorm ideas, explain concepts, and engage in back-and-forth dialogue. This is genuinely powerful technology, and it's the most accessible entry point to AI for most people. But it's one district, not the whole city.

Generative AI creates new content. This overlaps with conversational AI but extends far beyond it. AI now generates images from text descriptions, composes music, produces video, writes software code, and designs everything from molecules to marketing materials. The key characteristic is creation: these systems produce things that didn't exist before. This is the category driving much of the current excitement—and anxiety—about AI's impact on creative and knowledge work.

Agentic AI takes autonomous action toward goals. Rather than simply responding to human prompts, agentic systems can plan, execute multi-step tasks, use tools, and adapt their approach based on results. You might ask an agent to research competitors, and it will decide which sources to consult, what information to extract, and how to synthesize findings—without you specifying each step. These systems are newer and less mature than conversational AI, but they're advancing exceedingly fast. The shift from AI that responds to AI that acts is one of the most significant dynamics currently occurring.

Analytical AI focuses on finding patterns and making predictions. This is AI that examines data to surface insights humans might miss. Fraud detection systems that flag suspicious transactions. Demand forecasting tools that predict inventory needs. Medical imaging systems that identify tumors. Predictive maintenance algorithms that anticipate equipment failures. These systems often work invisibly, processing data in the background

20

and surfacing conclusions for human review. You may not interact with them directly, but they're increasingly shaping decisions across every industry.

Physical AI brings intelligence into the material world. This includes robots in warehouses and factories, autonomous vehicles navigating city streets, drones conducting inspections, and surgical systems assisting in operating rooms. Physical AI combines perception (understanding the environment through sensors), reasoning (deciding what to do), and actuation (taking physical action). It's the domain where AI meets atoms—and where the stakes often involve safety in ways that purely digital AI does not.

These five districts aren't walled off from each other. A warehouse robot uses analytical AI to optimize its routes, generative AI to adapt to new situations, and physical AI to navigate the space. A customer service system might combine conversational AI for interaction with analytical AI to predict customer needs. Powerful AI systems are increasingly multimodal—processing text, images, audio, and video together rather than treating them as separate inputs. The boundaries blur. But understanding the distinct capabilities helps you see the full scope of what AI can do—and spot opportunities you might otherwise miss.

How AI Thinks (And How It Doesn't)

Now let's talk about what makes this city fundamentally different from the software landscape you've navigated your whole career. This matters because the mental models that served you well for traditional technology can actively mislead you when applied to AI.

Traditional software is deductive. Programmers write explicit rules: if this, then that. The software applies those rules to inputs and produces outputs. Ask a traditional calculator what 2 + 2 equals, and it will always say 4. Ask it a million times, same answer. The behavior is deterministic—fully specified by the rules humans wrote. If something goes wrong, there's a bug in the code, and you can find and fix it.

AI is inductive. Instead of following rules that humans wrote, AI systems learn patterns from examples. Show an AI system millions of photos labeled "cat" and "dog," and it will learn to distinguish cats from dogs—without anyone programming explicit rules about whiskers or ear shapes. The system infers the pattern. It reasons from examples to general principles, rather than from general principles to specific cases.

This difference has profound implications.

First, AI can do things we don't know how to program. No one can write explicit rules for what makes a face look trustworthy, what makes prose flow well, or what makes a business strategy coherent. These involve tacit knowledge—patterns we recognize but can't fully articulate. Because AI learns patterns inductively rather than following explicit rules, it can capture this tacit knowledge in ways traditional software cannot. That's why AI can write passable poetry, recognize emotions in photographs, and generate plausible business analyses. We couldn't program these capabilities directly. We trained them.

Second, AI is probabilistic, not deterministic. Ask a traditional program the same question twice, you get the same answer. Ask an AI the same question twice, you might get different answers— both plausible, neither "wrong" in the way a software bug is wrong. This isn't a flaw; it's fundamental to how these systems work. They're generating probable responses based on patterns, not executing predetermined rules.

Sometimes the variation is a feature: you get fresh perspectives by asking again. Sometimes it's a challenge: you can't assume consistency without checking. Either way, it means AI requires different oversight than traditional software. You can't just test it once and trust it forever. You have to think in terms of probabilities, error rates, and acceptable ranges of variation. In a sense, this is more like dealing with people than with machines.

Third, AI fails differently than traditional software. When traditional software fails, it usually fails obviously—a crash, an error message, a clearly wrong output. When AI fails, it often fails subtly. It generates a confident-sounding answer that happens to be wrong. It produces analysis that's plausible but based on faulty assumptions. It hallucinates—presenting fabricated information as fact. These failures look like successes unless you know to check. This is why human oversight remains essential and why blindly trusting AI outputs is dangerous.

Fourth, AI reflects its training. The patterns AI learns come from the data it was trained on. If that data contains biases—and human-generated data inevitably does—the AI will learn those biases. If the training data is limited or unrepresentative, the AI's capabilities will be limited in corresponding ways. Understanding that AI is fundamentally shaped by its training helps you ask better questions: What data was this trained on? What perspectives might be overrepresented or missing? Where might this system's training not match our use case?

The shift from deductive to inductive thinking is one of the most important mental model updates you can make. AI is not a faster, smarter version of traditional software. It's a fundamentally different kind of tool—one that requires different expectations, different oversight, and different intuitions about where it will excel and where it will stumble.

What AI Can and Can't Do Reliably

So what can you count on? Where does AI deliver reliably, and where should you remain cautious?

AI excels at pattern matching at scale. The fraud detection that flags suspicious transactions, the system that predicts which machines need maintenance, the tool that identifies promising drug candidates—these all leverage AI's strength at finding patterns in vast amounts of data.

AI excels at generating variations. Need a hundred different headlines? A dozen approaches to explaining a concept? Multiple drafts to react to? AI is exceptional at producing variations on themes. Work that once forced you to pick a lane early can now start wide and narrow late.

AI excels at translation and transformation. Converting between formats—text to code, data to visualization, one language to another, verbose to concise—is a sweet spot for current AI systems. They're trained on countless examples of such transformations and can apply those patterns fluently.

AI excels at first drafts. Whether it's a document, a strategy, a piece of code, or an analysis, AI can produce a credible starting point much faster than a human working from scratch. Many people find that their best use of AI is as a drafting partner: AI produces the initial version, humans refine it.

Where should you remain cautious?

AI struggles with grounded novelty. AI systems can generate creative combinations and unexpected ideas at a pace no human team can match. The challenge is that many of these ideas aren't viable. They sound plausible but ignore physical constraints, regulatory realities, or practical limitations that experienced humans recognize instinctively.

AI systems can struggle with factual precision. They often state falsehoods with the same confidence as truths—hallucinating citations, inventing statistics, and fabricating quotes. Some systems are better than others at staying grounded, particularly when they're primed with curated content, but none are fully reliable.

AI struggles with knowing what it doesn't know. Humans can often sense when we're out of our depth—we feel uncertain, we hedge, we recognize the limits of our knowledge. AI systems typically lack this metacognition. They generate responses with consistent confidence whether they're in familiar territory or completely lost.

AI reasoning is improving fast—perhaps faster than any other capability. Recent systems can handle multi-step logic that would have stumped their predecessors by months, not years. But long chains of reasoning remain fragile. This is one area where the gap between AI and human judgment is closing quickly, which makes it worth reassessing regularly. The AI that couldn't handle your most complex reasoning last quarter might handle it this quarter.

TRY THIS

Ask an AI to explain something you're an expert in—a topic where you can evaluate the quality of its response. Where does it get things right? Where does it miss nuance or make subtle errors? This exercise builds calibrated trust: understanding where AI is reliable in your domain and where it falls short.

The Hype Filter

Understanding how AI works gives you a powerful filter for evaluating claims about AI capabilities. The landscape is full of vendors overpromising, startups exaggerating, and demos that don't represent production reality. A few conceptual tools can help you separate signal from noise.

Be skeptical if someone tells you their AI "solves" a problem without discussing accuracy rates, false positives, false negatives, and edge cases. Professionals talk about performance on specific benchmarks, error distributions, and where their systems struggle. Hype marketers talk about magic.

Be skeptical of demos without discussion of production realities. Demos are curated to show AI at its best. Production use means handling the worst cases, the edge cases, the cases no one anticipated. Ask what happens when the AI fails. Ask how often it fails. Ask what the recovery process looks like. The answers—or the inability to provide answers—tell you a lot about whether a capability is demo-ready or production-ready.

Be skeptical of claims that AI "understands." Current AI systems process patterns; whether they understand in any meaningful sense is a philosophical debate. Talk of understanding often obscures the real question: does the system perform reliably on the actual task, in the actual conditions, that matter to you?

Be skeptical of capabilities claimed for one domain but needed in another. An AI trained on general text won't automatically understand your specific industry, your proprietary data, your unusual use cases. Performance doesn't transfer automatically across domains. Always ask: has this been validated on data that resembles ours, with tasks that resemble ours, by users who resemble ours?

The goal isn't to become a cynic who dismisses all AI claims. Much of what AI can do is genuinely remarkable. The goal is to become a calibrated evaluator who can distinguish the remarkable from the exaggerated, the production-ready from the vaporware, the transformative from the merely trendy.

Compared to What?

When evaluating AI capabilities, many people make the same mistake: they compare AI to the perfect ideal.

"AI makes mistakes." Compared to what? A perfect expert who never errs? That expert doesn't exist—and if they did, you couldn't afford them.

"AI sometimes misses nuance." Compared to what? Your overstretched team working on their fourth coffee at 11 PM?

"AI lacks deep expertise." Compared to what? The consultant you can't get on the calendar for three weeks?

The relevant comparison isn't AI versus ideal human performance. It's AI versus your available alternative. That might be a junior employee still learning the ropes. It might be an expensive outside resource you use sparingly. It might be nothing at all—the task simply doesn't get done because no one has bandwidth.

When you frame the question realistically, the calculus often shifts dramatically. An AI that performs at the 70th percentile sounds mediocre compared to an expert. But if your actual alternative is an overworked generalist performing at the 40th percentile—or no one at all—suddenly 70th percentile looks transformative.

This doesn't mean lowering your standards. It means being honest about your baseline. "Compared to what?" is the question that cuts through both hype and unwarranted dismissal. It forces you to evaluate AI against the real world rather than an imaginary one.

Enough to Lead

You don't need to understand how AI works at an engineering level to lead effectively in the AI era. You don't need to know the mathematics of neural networks or the architecture of transformer models. Plenty of excellent executives will never train a model or write a line of code, and that's fine.

But you should have the mental map we've just walked through. You should know that AI is bigger than chatbots and that generative AI is inductive and probabilistic, not deductive and deterministic. You should to know where AI excels and where it stumbles, so you can deploy it where it's strong and protect against its weaknesses. And you need a hype filter to navigate a landscape where everyone is selling the future.

This foundation prepares you for what comes next. Because knowing about AI isn't the same as working with it. And that's where we turn now.

3

LEVELING UP WITH AI

There's an old saying in leadership: you can't steer a parked car. A car in motion—even if it's heading slightly wrong—can be turned, adjusted, corrected. A parked car just sits there. The same applies here. Leaders who wait until they fully understand AI before engaging with it will find themselves perpetually parked. The learning comes from the doing.

You can study a city's layout without ever walking its streets—and you can understand AI conceptually without ever developing the instincts that come from working with it. Intellectual understanding is necessary but not sufficient. At some point, you have to stop studying and start doing.

Now is the time to get in motion. We'll consider a framework for thinking about your AI journey, practical skills that will serve you at every stage, and a clear invitation to take the next step— whatever that step might be for you.

Level Up Framework

Not everyone starts from the same place, and not everyone needs to reach the same destination. But it helps to have a sense of the terrain—to know where you are, where you might go, and what becomes possible at each stage.

Think of AI fluency as a five-level progression. Each level represents a deeper integration of AI into how you think and work. The levels aren't fixed; some people will skip levels, others will move back and forth depending on context. But the framework helps you honestly assess where you are and identify concrete next steps.

One caveat: the floor keeps rising. What counted as Collaborator-level usage a year ago is Consumer-level today. The levels describe your relationship with AI, not a fixed set of skills. The specific capabilities at each level will keep advancing; the progression from tool user to strategic thinker remains constant.

Level 1: Consumer. At this level, you use AI as a tool for discrete tasks. You might ask ChatGPT to draft an email, summarize a document, or explain a concept. The interaction is transactional—you have a task, AI helps with it, you move on. Most executives who've experimented with AI are at least here. It's a valid starting point, but it barely scratches the surface of what's possible. At this level, AI saves you time on individual tasks, but it doesn't fundamentally change how you think or work.

Level 2: Collaborator. Here, AI becomes a thinking partner rather than just a task-completer. You don't just ask AI to draft an email; you ask it to help you think through the communication strategy first. You use AI to pressure-test your ideas, explore alternatives, and challenge your assumptions. The relationship shifts from "do this for me" to "think with me." This is where AI starts to genuinely augment your judgment rather than just your productivity. What changes: you start having better ideas, not just faster execution.

Level 3: Integrator. At this level, AI becomes embedded in workflows rather than being something you turn to occasionally. You've identified the recurring tasks and decisions where AI adds value, and you've built habits around using it. Maybe you start every strategic analysis with an AI brainstorm. Maybe you have an AI agent review every document before finalizing it. The key shift is from episodic use to systematic integration—AI is essential to how you work, not an occasional supplement. What changes: your baseline capabilities expand, and you start to forget what work felt like without AI.

Level 4: Orchestrator. Now you're deploying AI across teams and processes you lead. You're thinking about how AI can enhance your organizational capabilities, not just individual effectiveness. This might mean implementing AI tools for your team, deploying

AI agents that handle cross-functional organizational tasks, redesigning workflows to incorporate AI assistance, or developing proactive guidelines for how people should use AI. The focus expands from individual productivity to organizational capability. What changes: you become a multiplier, helping others level up while continuing your own journey.

Level 5: Transformer. This level is different from the others. It's less a skill to develop and more a perspective that emerges once you've internalized the first four. At some point, you stop asking "how do I use AI better?" and start asking "what should we be doing differently?" What business models become viable when intelligence is abundant? What assumptions about your industry might AI invalidate? This is the shift from doing things differently to doing different things—and it's where the real strategic value lives. You can't force your way here. But the fluency you build at the other four levels is what makes it possible to see opportunities that others miss.

Where are you right now? Be honest. Most executives I talk to are somewhere between Consumer and Collaborator—they've used AI for tasks but haven't yet made it a true thinking partner or integrated it into their regular workflows. There's no shame in that; everyone starts somewhere. The question is: what would it take to move up one level?

Skills That Help You Level Up

Moving up the levels isn't just about using AI more often—it requires developing specific capabilities. Start with three: effective prompting, building your AI consulting team, and working with AI agents.

Effective prompting is the foundation. How you communicate with AI dramatically affects what you get back. This isn't about memorizing magic formulas—it's about understanding principles that make your interactions more effective.

Four questions shape the difference between a mediocre AI interaction and a genuinely useful one:

What does the AI need to know? Context matters enormously. Who are you? What's the situation? What background would a smart colleague need to help you well?

What are you trying to accomplish? This isn't as obvious as it sounds. "Write me an email" is a task. "Convince my skeptical CFO to fund a pilot program" is an intent. When you share the underlying goal, AI delivers better results.

What perspective would be most helpful? You can ask AI to think like a skeptical investor, an enthusiastic customer, a cautious attorney, or a creative strategist. Each perspective surfaces different considerations. This is one of the most powerful and underutilized prompting techniques—AI doesn't have a natural "self" the way you do, so it can genuinely adopt different frames of thinking.

What should the output look like? Format, length, tone, what to include, what to avoid. A one-paragraph summary or a detailed analysis? Ready to share with your board or just for your own thinking? Constraints shape quality.

Iteration is normal. Your first prompt rarely produces the perfect result—and that's fine. Treat AI interaction as a conversation, not a single query. Refine, redirect, ask follow-ups. "That's good, but make it more concise" or "Now give me the counterarguments" or "What am I missing?" The best results often come from the third or fourth exchange, not the first.

Personalization compounds your investment. Most AI platforms let you configure persistent context—your role, your industry, your communication preferences—so every conversation doesn't start from zero. Some let you attach project files that ground the AI in your specific situation. Others develop memory across conversations, learning your priorities and patterns over time. Setting up this context is one of the highest-leverage moves you can make—it's the difference between working with a different intern every day and working with a colleague who knows your business.

TRY THIS

Pick a decision you're facing—something substantive, not trivial. Spend fifteen minutes working through it with AI. Start by giving context: what's the decision, what are your options, what factors matter most, what constraints exist. Then ask AI to help you think through it. This single exercise will teach you more about AI collaboration than any amount of reading.

Next, building your **AI consulting team** accelerates your journey from Consumer to Collaborator. Instead of using AI as a single generic assistant, create distinct personas that serve different roles in your thinking.

Set up three AI conversations, each with a different brief. The first is your supportive advisor — a strategic thought partner who helps you develop ideas, build on your thinking, and spot opportunities you're missing. The second is your skeptical CFO — wary of hype, focused on concrete returns, primed to challenge your assumptions and probe for weaknesses. The third is your key customer — someone evaluating whether what you're proposing truly solves a problem they care about, at a price they'd pay, in a way that fits how they work.

TRY THIS

Take a real initiative or proposal you're working on and run it through multiple personas such as supportive advisor, skeptical CFO, and key customer. Notice how each perspective surfaces different considerations. For a more advanced exercise, describe all three perspectives in a single conversation and ask the AI to simulate a discussion among them while you observe.

Running your ideas through all three gives you something close to a personal advisory board — different perspectives that strengthen your thinking before you take it to real stakeholders. Can we build this? Can we afford this? Does anyone want this? It's a concrete way to use AI as a thinking partner rather than just a task-completer, and it's available to you right now, today, with any conversational AI tool. When these advisors start executing, not just advising, you'll enter the world of AI agents

While your AI consulting team responds to prompts and advises, **AI agents** take autonomous action toward goals. You give an agent an objective—research these competitors, analyze these documents, monitor these metrics and flag anomalies—and it figures out the steps, uses tools, and delivers results without you specifying every move.

This is a significant shift in how you work with AI. Instead of driving every interaction, you're delegating. Instead of thinking "what should I ask AI to do next," you're thinking "what outcome do I want, and can I trust AI to figure out how to get there?" Agent capabilities are evolving rapidly, managing workflows, accessing your files, and coordinating tasks—functioning less like tools and more like trusted team members.

TRY THIS

Practice agent delegation. Choose a task with clear deliverables—competitive briefing, market analysis, crafting a presentation. Define the specific outcome you want (not just the activity, but what "done" looks like) and the boundaries (what sources or tools should the agent use, and which should it avoid). Then delegate the task to an AI agent and observe: Where did it need more guidance? Where did it surprise you with capabilities you didn't expect? Where did it go off track?

For leaders, this means developing new skills around delegation and oversight—defining objectives clearly enough that an agent can work autonomously, then verifying that the results meet your standards. The executives who learn to work effectively with agents will have a significant advantage over those who remain stuck in purely conversational AI use.

Climbing vs. Leaping

The Level Up framework suggests a natural progression—climb the levels one by one, building competence at each stage before advancing to the next. And often, that's good advice. You develop

34

instincts at Level 2 that serve you at Level 3. You learn what AI can and can't do through hands-on experience that no amount of reading can replace.

But sometimes the right move is to leap, not climb.

Consider the difference between iterating on a better iron lung versus inventing the polio vaccine. The iron lung was a sophisticated response to polio—a machine that helped patients breathe when the disease paralyzed their respiratory muscles. Engineers kept improving it: better materials, more reliable mechanics, greater patient comfort. Incremental progress, each step building on the last. But the polio vaccine didn't iterate on the iron lung. It leaped past it entirely by solving the problem at a different level. Once the vaccine existed, better iron lungs became irrelevant.

Sometimes your situation calls for climbing: steady progression, building capabilities, moving up one level at a time. And sometimes it calls for leaping: skipping intermediate stages because a fundamentally different approach is available. The executives who thrive will be those who can recognize which situation they're in. Are you in an iron-lung moment, where incremental improvement is the right path? Or a vaccine moment, where a leap to a different level makes more sense?

Either way, the key is to start moving. A parked car can't be steered, even toward a breakthrough.

The Case for Starting Now

If you've been waiting for AI to mature before engaging seriously, I want to challenge that instinct. Yes, AI is evolving rapidly. Yes, the tools available today will be surpassed by better tools tomorrow. But the executives who wait for stability will be waiting forever—and falling further behind with each passing month.

The skills that matter most aren't tool-specific. Knowing which buttons to click in a particular interface is the least durable kind of knowledge. But knowing how to frame problems for AI assistance, how to evaluate AI outputs critically, and how to integrate AI into your decision-making are meta-skills that transfer across tools and will remain valuable regardless of which specific systems dominate. Moreover, fluency compounds. The executives who start building AI fluency now will develop intuitions that accelerate their learning as new capabilities emerge.

There's a leadership dimension as well. Your people are watching. If you're visibly engaging with AI—experimenting, learning, integrating it into your work—you signal that AI matters and that it's safe to explore. The executives who lead AI transformation most effectively are those who've done the personal work first.

Where to Start

Wherever you are in the Level Up framework, there's a concrete next step available to you.

If you're at Level 1 (Consumer), your next step is to shift from task completion to thinking partnership. Pick a real challenge you're working on—not a test case, but something that actually matters—and use AI to think it through, not just execute it. Try the AI consulting team technique. Experience what it feels like to have AI augment your judgment, not just your productivity.

If you're at Level 2 (Collaborator), your next step is integration. Identify one recurring workflow where AI could add value—whether through direct interaction or an AI agent handling it systematically—and commit to using it consistently for a month. Build the habit. Notice what works and what doesn't. Let the integration become automatic rather than effortful.

If you're at Level 3 (Integrator), your next step is to expand your scope. Where could AI agents handle cross-functional tasks your team currently does manually? What processes could be redesigned with AI assistance? Start one pilot. Learn from it. Then expand.

If you're at Level 4 (Orchestrator), your next step is to look up from operations. Step back from operational deployment and ask bigger questions. What does AI abundance mean for your business model? What assumptions might it invalidate? What opportunities might it create? Give yourself permission to think at the Transformer level, even if you're not ready to act there yet.

And if you're already at Level 5 (Transformer)—if you're actively rethinking your business through an AI-abundance lens—your work is just beginning.

Getting in Motion

While you can't steer a parked car, once you're in motion, turning becomes possible.

You don't need to have AI figured out before you start. You don't need to reach Level 5 before you can lead effectively. You don't need to master every tool or anticipate every development. What you need is to grab your mental map and get moving—to engage with AI seriously enough to develop real fluency, to climb or leap as the situation demands, to build the instincts that only come from experience.

That foundation prepares you for what comes next: thinking strategically about how AI reshapes competition, business models, and the nature of work. Part II takes everything you've learned about seeing clearly and applies it to the strategic questions that will define your organization's future.

But before you turn the page, consider: What's your next step?

Part II

THINKING STRATEGICALLY

AI changes the rules of competition, the logic of business

models, and the nature of work.

4

HOW AI RESHAPES COMPETITION

Getty Images and Shutterstock spent decades building the most comprehensive visual libraries in the world. Hundreds of millions of photographs, each one licensed, tagged, and legally cleared. The archive was the moat—no competitor could replicate that depth overnight. Then generative AI arrived. Adobe's Firefly image generator produced three billion images within months of launch, surpassing the total archives of most traditional photo libraries. After precipitous declines, the two companies announced a planned merger. Perhaps the most telling detail: Shutterstock earns over a hundred million dollars licensing its image data to the very AI companies displacing it—selling raw material for the tools that are making its core business obsolete.

Stories like this are multiplying across industries. Advantages that took decades to build are being neutralized in months. Barriers that seemed permanent are dissolving. And the executives who built those advantages are watching, trying to figure out what just happened.

If you've been in business long enough, you've internalized certain truths about competition. Industries have structures. Some positions are stronger than others. Advantages, once built, tend to persist. These truths aren't wrong, exactly—but AI is stress-testing every one of them.

The Five Forces, Revisited

Michael Porter's Five Forces framework has shaped competitive strategy for over four decades. It remains useful—not because it's sacred, but because it asks the right questions. What determines the intensity of competition in your industry? What affects your

41

power relative to suppliers and customers? Where do threats come from?

AI is changing the answers to all of these questions. Let's walk through each force, noting where AI's impact is most pronounced.

Threat of new entrants. Barriers to entry are shifting in contradictory directions. On one hand, AI lowers many traditional barriers. Capabilities that once required large teams and years of development can now be assembled quickly using AI tools. A startup can build a sophisticated product with a fraction of the headcount that would have been required five years ago. The capital requirements for launching a software business have plummeted. This means incumbents face potential challengers they never would have seen before—small teams moving fast, unburdened by legacy systems or organizational inertia.

On the other hand, AI may be creating new barriers at the top. Companies with massive data sets, computational resources, and AI talent may be pulling away from everyone else. If the best AI models require billions in training costs and proprietary data that took decades to accumulate, the barrier to competing at the frontier may be higher than ever. The startup that can quickly build a good product may find it impossible to build a great one.

The net effect depends on your industry. In some sectors, AI is democratizing capability and inviting new entrants. In others, it's concentrating power among those who got there first. The strategic question: which dynamic dominates in your market?

Bargaining power of suppliers. AI's impact on supplier power varies by what's being supplied. For commodity inputs— standardized components, generic services, interchangeable resources—AI often shifts power toward buyers. Better analytics mean better price discovery. Automated procurement means easier supplier switching. Predictive systems reduce dependency on any single supplier.

But for suppliers of AI capabilities themselves—training data, specialized models, computational infrastructure, scarce AI talent—the power dynamic may favor sellers. If your AI strategy depends on a small number of infrastructure or model providers, you've introduced a new supplier dependency that didn't exist before. Some companies are discovering that their AI ambitions are constrained by suppliers they didn't anticipate needing.

The strategic question: are your key suppliers becoming more interchangeable or less? And have you inadvertently created new supplier dependencies in pursuit of AI capabilities?

Bargaining power of buyers. AI is generally empowering buyers—customers can now compare offerings more easily, evaluate alternatives more rigorously, and switch providers with less friction. AI tools help procurement teams analyze contracts, identify hidden costs, and negotiate more effectively. The information asymmetries that once favored sellers are collapsing.

For business-to-consumer companies, AI-powered comparison shopping, reviews, and recommendations make it harder to win on anything other than genuine value. For business-to-business companies, sophisticated buyers are using AI to evaluate vendors with a rigor that was previously impractical. They're benchmarking your performance, modeling alternatives, and identifying switching opportunities.

The strategic question: what happens when your customers get as good at analyzing you as you are at analyzing them?

Threat of substitutes. This is where AI's impact may be most dramatic—and most precarious for incumbents. AI doesn't just enable existing competitors to do what they already do better. It enables entirely new approaches that substitute for what you offer.

Consider manufacturing. Traditional competitive analysis might focus on rival factories—their capacity, their costs, their quality. But the real threat might be a distributed network of AI-optimized micro-factories that can produce customized goods closer to end customers, eliminating the logistics advantages that centralized manufacturing once enjoyed.

The danger of substitution is that it often comes from outside your traditional competitive set. You're watching your known competitors while the real threat emerges from a different

43

direction entirely. The startup that replaces you won't look like you, won't market to your customers through familiar channels, and won't show up in your competitive analysis until it's too late.

The strategic question: what do your customers really want, and could AI enable them to get it without you?

Competitive rivalry. AI is intensifying competitive rivalry in most industries. When AI tools are available to everyone, they stop being a source of advantage and become table stakes. The productivity gains you achieve, your competitors achieve too. The insights you generate from data, they generate as well. AI becomes the cost of competing, not the means of winning.

This creates pressure toward commoditization. When everyone can produce sophisticated analysis, analysis stops being a differentiator. When everyone can generate professional content, content stops being a differentiator. When everyone can personalize at scale, personalization stops being a differentiator. What's left to compete on?

ASK AI

Ask an AI to analyze your industry using Porter's Five Forces, then follow up: "How might AI shift each of these forces over the next three years? Where does AI strengthen our position, and where does it weaken it?" The analysis won't be perfect, but it will surface questions worth asking.

Data Advantage and the Moat Mirage

Much has been made of "AI moats"—the idea that certain companies have insurmountable advantages because of their data. The logic seems compelling: more data leads to better AI models, which attract more users, which generate more data. A virtuous cycle that compounds over time, leaving competitors permanently behind.

There's truth to this. Scale advantages in AI are real. A recommendation system trained on billions of interactions will generally outperform one trained on millions. A language model trained on the entire internet has capabilities that a smaller model can't match. Data advantages do exist, and they do matter.

But the moat narrative is often oversimplified, and executives who rely on it may be dangerously complacent.

First, not all data advantages are created equal. The question isn't just "how much data do you have?" but "is your data defensible, relevant, and continuously refreshed?" Proprietary data that no one else can access may be valuable; data that competitors can acquire or approximate is less so. Historical data depreciates as the world changes; a decade of consumer behavior data may tell you less than you think if buying patterns have recently shifted.

Second, foundation models are changing the game. When powerful pre-trained models become widely available—through open-source releases or API access—companies can achieve sophisticated AI capabilities without building from scratch. The startup that would have needed years of data collection to compete can now fine-tune an existing model on a much smaller dataset. This doesn't eliminate data advantages, but it compresses them.

Third, data advantages can be disrupted by architectural shifts. Blockbuster had unmatched data on rental patterns, regional preferences, and inventory optimization across thousands of locations. None of it mattered when Netflix changed the delivery model entirely. Similarly, advances in AI techniques—synthetic data, transfer learning, more efficient training methods—can undermine advantages that seemed permanent. Today's insurmountable data moat may be tomorrow's legacy asset.

The honest assessment: data advantages matter, but they're narrower and more fragile than the moat narrative suggests. If your competitive strategy depends on data advantages, pressure-test it. Ask: What would it take for a competitor to achieve "good enough" AI capabilities without our data? How quickly is our data depreciating? Could an architectural shift make our data advantage irrelevant?

Strategic Postures for an AI Era

The moat metaphor—an advantage so deep competitors can't cross it—may be the wrong frame entirely for many companies in an AI-disrupted landscape. Not every company needs a moat. Not every situation rewards moat-building. And clinging to a moat mentality can blind you to better strategic options.

Consider some alternatives.

Agility over position. In fast-moving environments, the ability to learn and adapt quickly may matter more than any defensible position. By the time you've built a moat, the river may have moved. Some companies thrive not by building walls but by staying in motion—fast followers who don't need to be first because they can catch up quickly, or nimble innovators who abandon positions as readily as they claim them.

Niches over scale. Not every valuable business needs to be a platform. AI may enable smaller companies to serve specialized needs with a precision and efficiency that wasn't previously viable. The global platform captures headlines; the focused niche captures margins. Sometimes the right response to AI-driven commoditization is to go narrower, not broader.

Partnerships over ownership. Since AI capabilities are increasingly available through APIs and partnerships, building everything yourself may be the wrong strategy. The company that assembles best-in-class capabilities from multiple sources may outperform the one that insists on owning every piece. Competitive advantage shifts from what you own to how you orchestrate.

Relationships over transactions. As AI commoditizes the transactional elements of business, the relational elements become more valuable by contrast. The company that can't be easily replaced because of deep customer relationships, ecosystem embeddedness, or institutional trust may have something more durable than any technological moat.

The point isn't that moats don't matter. The point is that "build a moat" isn't a universal strategy. The right strategic posture

46

depends on your situation—your starting position, your capabilities, your industry dynamics, and your honest assessment of what's defensible.

ASK AI

"Play a startup trying to disrupt my company using AI. You have access to modern AI tools and foundation models, but none of our proprietary data or existing customer relationships. How would you attack our position? What would you build? Who would you target first?" The answers reveal your actual vulnerabilities.

Diagnosing Your Competitive Position

Rather than prescribing a single strategy, let me offer a diagnostic framework. The right response to AI-driven competitive shifts depends on where you sit.

If your current position is strengthening: Some companies find that AI amplifies their existing advantages. They have the data, the talent, the distribution, and the customer relationships to benefit disproportionately from AI capabilities. If this is you, the strategy is acceleration—invest aggressively, widen the gap, and don't let challengers close the distance.

If your current position is eroding: Other companies find that AI is dissolving the advantages they've built over decades. The expertise they've accumulated is being commoditized. The barriers that protected them are falling. The value they provide is now available cheaper or faster elsewhere. If this is you, the strategy is reinvention—find the new basis for value before the old one collapses. This is a "yes, and" moment: yes, the old advantages are fading, and here's where we build the new ones.

If your position is stable but vulnerable: Many companies fall in the middle. AI hasn't disrupted them yet, but they can see it coming. The current business still works, but the trajectory is concerning. If this is you, the strategy is managed transition—

47

maintain the current business while building the capabilities and positions you'll need in the future.

If you're a potential disruptor: And some readers are on the other side of this equation—in positions to challenge incumbents using AI. If this is you, the strategy is targeted attack—identify where AI creates asymmetric opportunities, where incumbents are slow or complacent, where customer needs are going unmet.

Honest diagnosis is the prerequisite for sound strategy. The executives who get this wrong convince themselves their position is stronger than it is, that disruption is further away than it is, that their moat is deeper than it is. The diagnosis has to be unflinching.

The View from the Customer

Competitive strategy ultimately comes back to the customer, and AI is giving customers new options. The competitive landscape isn't just being reshaped by what companies do with AI. It's being reshaped by what customers can do with AI themselves.

The companies that win will be those that understand most clearly what value they provide that AI cannot—and build their positions around that.

Your competitive moat may be eroding faster than you think. Or it may be deeper than it's ever been. The first step is diagnosing which situation you're in. The second step is having the courage to act on what you see.

Now that we've examined how AI reshapes the competitive landscape, we need to go deeper. Competition is about relative position, but business models are about how you create and capture value in the first place. That's where we turn next.

5

BUSINESS MODELS FOR ABUNDANCE

We've seen how AI reshapes competition, but that's only part of the strategic picture. When the underlying economics of an industry shift, business models that worked beautifully can become liabilities overnight. The strategies that made you successful—the pricing structures, the delivery mechanisms, the assumptions about what customers will pay for—may be precisely what prevents you from adapting.

AI as Business Model Enabler

The most obvious impact of AI on business models is efficiency: doing the same thing faster or cheaper. But the more interesting impact is enabling business models that weren't previously viable.

Consider law firms. They've traditionally billed by the hour because legal work required expensive human attention for every task. But AI is making much of this work abundant. Document review that once required teams of junior associates can now be accomplished in hours. Contract analysis that billed at hundreds of dollars per hour can be performed for pennies per page.

Cost reduction alone misses a bigger impact—changing what business models become possible. Some firms are shifting to fixed-fee arrangements, where profit comes from efficiency rather than duration. Others are moving toward success-based pricing, tying fees to outcomes rather than inputs. The AI capability enables business models that would have been financially reckless before.

Similar transformations are happening across industries—consulting, financial services, software, healthcare. The pattern is consistent: AI makes certain inputs abundant, which destroys

49

business models built on selling those inputs. But what replaces them isn't always the same.

Some businesses will shift to outcome-based models—fixed fees, success-based pricing, pay-for-performance. This works when results are measurable and clients care about the destination more than the journey. But others will find their value lies in the process itself—the trust, the relationship, the experience of working with someone who understands your situation. Still others will compete on judgment: not producing the answer, but knowing which question to ask and what to do with the result.

The question for any business: what are your customers really paying for? The outcome, the experience, the judgment, or something else entirely? Your business model needs to be built around the honest answer.

The Cannibal's Dilemma

Here's the uncomfortable reality: the business model changes AI enables often threaten the business models that made companies successful in the first place. This is Clayton Christensen's innovator's dilemma, applied to the AI moment.

Christensen's framework is often misunderstood. The companies that fail to adapt aren't stupid—they're successful precisely because they've optimized beautifully for the current reality. A law firm that has spent decades building a profitable practice around billable hours has good reasons to resist fixed-fee pricing: partner compensation depends on billable hours, associate development assumes years of billable work, client relationships are built around monthly invoices. Everything about the organization is optimized for the current model. This isn't a failure of vision—it's the natural consequence of success.

The dilemma is real: cannibalize your existing business model before someone else does, sacrificing near-term profits, or wait and risk being disrupted by competitors unencumbered by your success.

But here's what the disruption narrative often misses: sometimes the right answer is to defend the existing model, at least for a while. Not every business needs to race toward AI-enabled transformation. Some markets will shift slowly. Some customers will pay premiums for the old way. Some competitive positions are more defensible than they appear. The real question is: "how fast is our world shifting, and what's our realistic window?"

The Renovation Problem

Here's an uncomfortable truth about AI transformation: it's much easier to build new than to renovate existing.

Anyone who's worked on a house knows this. You can't just swap out the plumbing while people are showering. You discover that the walls you need to move are load-bearing. The electrical system that seemed fine turns out to be incompatible with modern appliances. Every fix reveals three more problems.

Organizations work the same way. Retrofitting AI into existing operations means working around legacy systems, established processes, incumbent incentives, and organizational muscle memory. Every integration point is a potential failure point. Every workflow change disrupts people who were doing just fine before you showed up.

This is why startups are dangerous. They're not renovating; they're building greenfield. No legacy systems to integrate with. No existing processes to preserve. No organizational antibodies attacking the new approach. They can design AI-native from day one, while you're trying to retrofit AI into systems designed for a different era.

The asymmetry is stark: your competitor builds in months what takes you years. Not because they're smarter, but because they're not dragging the weight of your success behind them.

Thankfully, Christensen's innovator's dilemma isn't just a diagnosis—it comes with a prescription. When your existing business creates structural resistance to necessary change, you have three options.

Option 1: Retrofit. Transform your existing operations. Integrate AI into current systems, retrain current people, evolve current processes. This is the default approach, and sometimes it's the right one—especially when your competitive position is strong enough to absorb the J-curve, when the transformation can be incremental, or when the alternative would abandon assets and relationships that remain genuinely valuable.

The risk: you move slowly while competitors move fast. The renovation never quite finishes. The organization's immune system keeps rejecting the transplant.

Option 2: Parallel Build. Create new AI-native capabilities alongside your legacy operations. Run both in parallel. Let the new system prove itself before migrating customers or sunsetting the old. This gives you the greenfield advantage without abandoning what works—but it also means running two systems, two cost structures, and potentially two cultures.

The risk: the parallel build gets starved for resources because the legacy system still pays the bills. Internal competition between old and new creates political warfare. You never actually make the transition.

Option 3: Separate Entity. Spin out or create a distinct organization to pursue the AI-native approach—different leadership, different incentives, different culture, potentially different brand. This is Christensen's classic prescription: protect the new venture from the parent organization's antibodies by giving it structural independence.

The risk: you're now competing with yourself. The separate entity may succeed at cannibalizing your core business faster than external competitors would have. And organizational separation doesn't guarantee success—plenty of corporate ventures fail despite structural independence.

There's no universally right answer. The choice depends on how fast your market is shifting, how strong your current position is, how much organizational capacity for change you have, and how willing you are to cannibalize yourself before someone else does.

Retrofit makes sense when your core business has runway, when the AI transformation can be incremental, and when your organizational culture can absorb sustained change.

Parallel build makes sense when you need the greenfield advantage but can't afford to abandon existing operations—and when you have the resources and discipline to genuinely support both.

Separate entity makes sense when the required changes are so fundamental that they can't survive inside the existing organization—when the antibodies are too strong, the legacy incentives too entrenched, the cultural gap too wide.

What doesn't work is pretending the renovation problem doesn't exist—assuming you can transform at startup speed while carrying incumbent weight. The organizations that navigate AI transformation successfully are the ones that honestly assess how much their existing structures help versus hinder, and choose their approach accordingly.

A Strategic Choice

As AI makes certain capabilities abundant, businesses face a fundamental strategic choice: race toward abundance or defend scarcity.

The abundance strategy embraces AI-enabled scale. If AI can produce something at near-zero marginal cost, produce a lot of it. Flood the market. Win on volume, on reach, on the long tail of needs that weren't worth serving when production was expensive.

The scarcity strategy goes the opposite direction. As AI commoditizes the common, differentiate on the rare. Position your offering as the premium alternative—human-crafted, carefully curated, irreducibly personal. Command higher prices from customers who value what AI can't easily replicate.

Either strategy can work. Neither is inherently superior. The right choice depends on your market, your capabilities, and a sober assessment of what customers truly value.

The danger is getting stuck in the middle—neither achieving the scale advantages of abundance nor commanding the premium of genuine scarcity. The mushy middle, where you're using AI but not fully, charging less than premium but more than commodity, serving some customers adequately but none exceptionally. This is where competitive advantage goes to die.

The Authenticity Trap

When something becomes abundant, people search for new ways to signal value. One common response is to reject the abundant thing entirely and claim authenticity through its absence. "No AI was used" becomes a badge of honor—a kind of "organic" label for knowledge work as a response to AI slop.

This impulse is understandable. In a world of synthetic content, genuine human creation feels precious. In a world of algorithmic optimization, human idiosyncrasy feels refreshing. But the authenticity framing often rests on fuzzy thinking, and it's worth examining carefully.

Consider a theater company staging Hamlet. No one expects them to have written the play. The value isn't in the script's originality— it's in the interpretation, the performance, the production choices. External inputs are welcomed as legitimate because everyone understands where the contribution lies.

Consider a professor assigning a textbook. No one accuses them of inauthenticity for not writing the book themselves. The value isn't in the textbook—it's in the selection, the contextualization, the discussion, the mentorship. External inputs are legitimate because everyone understands what value the professor is providing.

Consider an accountant using financial modeling software. No one suggests they should calculate spreadsheets by hand to be authentic. The value isn't in the arithmetic—it's in the assumptions, the strategic interpretation, the judgment about what the numbers mean.

Yet when AI enters the picture, this logic somehow breaks down. A writer who uses AI to draft and then extensively revises is suspected of inauthenticity. A consultant who uses AI to generate initial analyses and then applies expert judgment is somehow cheating. A designer who uses AI to explore variations and then selects and refines the best is not a "real" designer.

Authenticity isn't about whether AI touched something. It's about whether the value you're claiming is honestly represented.

If you're selling "handcrafted artisanal content" and you're actually running prompts through ChatGPT with minimal editing, that's inauthentic—not because AI was involved, but because you're misrepresenting what you're providing. If you're selling strategic insight and you're using AI to accelerate research, generate options, and draft deliverables while you provide the judgment, direction, and accountability—that's perfectly authentic. The value you're claiming is the value you're providing.

The consultant who uses AI to accelerate analysis and produce polished deliverables faster isn't being inauthentic. They're being efficient. Their value was never "I can make slides slowly"—it was "I can solve your problem."

Authenticity, properly understood, is about honest representation of value. And here's the uncomfortable implication: if your definition of authenticity requires inefficiency, you're not protecting quality—you're protecting status.

That's not necessarily wrong. Status has real value. There are markets for things that are valuable precisely because they're difficult, scarce, or inefficient. A hand-written letter means something a printed one doesn't. There's no shame in competing on these dimensions.

But be honest about what you're selling. If the value you provide is genuinely your human effort, attention, or craft—sell that. If the value you provide is the outcome, the insight, the result—then refusing to use tools that would deliver better outcomes isn't protecting authenticity. It's protecting a story you tell yourself about your own importance.

And be careful about conflating natural with good. Poison ivy is all natural. That doesn't make it good for you. "No AI used" is not an automatic marker of quality. It's just a description of process. The quality of the output is what matters.

Value Proposition Revisited

The authenticity trap is ultimately the value proposition question in disguise: if AI can do this, why do customers need us?

For organizations facing this question, the authenticity debate often functions as avoidance. It's easier to stake out a position of principled resistance than to honestly examine whether your value proposition survives contact with abundance.

But avoidance isn't strategy. Eventually, the question demands an answer. This is another "yes, and" moment: yes, AI can do much of what we used to do, and here's what we provide that AI cannot.

Some organizations will find that their human premium is genuinely defensible. There are things customers will pay more for specifically because humans did them—not as a process fetish, but because the human involvement creates real value. Custom furniture. Bespoke tailoring. Personalized advice from someone who knows your situation deeply. In these cases, the "no AI" position isn't a trap—it's a legitimate strategic choice.

Other organizations will find that their human premium is sadly indefensible—a story they've been telling themselves while customers quietly calculate whether the premium is worth it. The chain bookstore insisting that people prefer to browse in person. The taxi company insisting that people prefer human dispatchers. Each of these beliefs was true for some customers, some of the time, but not enough to sustain the business model once better alternatives emerged.

The biggest opportunities in the AI era aren't efficiency gains within existing business models. They're fundamental rethinking of what business you're in. It's the difference between doing things differently and doing different things.

Doing things differently means using AI to improve your current operations—faster, cheaper, more accurate versions of what you already do. These improvements are valuable, but they're also defensive—they help you keep pace as AI raises the baseline for everyone.

Doing different things means using AI to pursue opportunities that weren't viable before. New products. New services. New business models. Markets you couldn't serve. Value you couldn't create. This is where transformation happens—and where lasting competitive advantage lives.

The distinction matters because efficiency gains are quickly competed away. If AI lets you produce reports 50% faster, that's an advantage—until your competitors adopt the same tools and match your speed. Then you're back to parity, just at a higher baseline

---

ASK AI

"What would an AI-native competitor in my industry look like if they started from scratch today, with no legacy business model to protect? What would they charge for? How would they price it? What would they not bother doing at all?"

The question isn't how AI can improve your current business. It's what business you should be in—and whether AI changes the answer.

Business model questions inevitably lead to questions about people. If AI changes what's valuable, it changes which human capabilities matter. If AI enables new business models, it requires new organizational structures to deliver them. If authenticity is about the human contribution, we need to understand what humans contribute that AI cannot.

That's where we turn next: to people, talent, and the changing nature of work.

6

THE HUMAN PREMIUM

AI reshapes competition and business models, but strategies don't execute themselves and organizations don't spontaneously transform. Behind every competitive position and value proposition are people—the humans whose capabilities, decisions, and efforts make organizations work.

AI is changing what it means to be one of those people.

This isn't primarily a story about job losses, though that narrative dominates the headlines. It's a more interesting and more nuanced story about transformation—about which human capabilities become more valuable, which become less, and how the nature of work itself is shifting. Leaders who understand this transformation can navigate it in ways that strengthen their organizations and honor the people who built them. Leaders who don't will find themselves managing fear and confusion rather than leading change.

How can we understand different ways humans and AI work together? What about the emergence of AI teammates? And then there's the individual version of the value proposition question: what are you contributing, and does AI change the answer?

Four A's of Human-AI Collaboration

The debate over whether AI will "replace jobs" misses a fundamental point: a job isn't a single thing. It's a bundle of tasks, and AI affects different tasks differently.

Consider a financial analyst. The job title sounds like one role, but it's actually a bundle: gathering data, cleaning and organizing information, running calculations, identifying patterns, building models, interpreting results, communicating findings, building relationships with stakeholders, exercising judgment about what matters. AI is already transforming some of these tasks dramatically while barely touching others.

When you view jobs as monolithic units, AI's impact seems binary—either the job survives or it doesn't. But when you view jobs as task bundles, you see a more nuanced reality: tasks are being automated, augmented, or untouched at different rates, and jobs are being re-bundled around the tasks that remain distinctly human.

The unit of analysis for AI deployment isn't the job—it's the task. Successful implementation requires decomposing work into its component tasks and addressing each appropriately

The Four A's framework provides a map for these differences. It identifies four distinct modes of human-AI collaboration, organized along two dimensions: task ambiguity and AI control. The combination determines which mode fits your situation.

	LOW AI CONTROL	HIGH AI CONTROL
HIGH TASK AMBIGUITY	AUGMENT *Human leads, AI assists*	AGENT *AI acts within human-set goals*
LOW TASK AMBIGUITY	ANALYZE *AI surfaces insights for review*	AUTOMATE *AI handles end-to-end*

Task ambiguity asks: how well-defined is what we're trying to accomplish? Some tasks have clear parameters and established patterns—the "right answer" is knowable. Others are novel, complex, or require judgment—the path forward is uncertain.

AI control asks: who's navigating the work? In some modes, humans remain firmly in charge, with AI supporting their decisions. In others, AI takes the wheel, acting autonomously within defined boundaries.

Let's walk through each quadrant.

Augment is the mode most people are familiar with. The task is ambiguous—novel, complex, or requiring judgment—and the human stays in control. You're writing a strategy document, and AI helps you brainstorm, draft sections, or pressure-test your thinking. You're preparing for a negotiation, and AI helps you anticipate counterarguments. The human remains in charge throughout—setting direction, evaluating outputs, making final calls. AI amplifies your capabilities without replacing your judgment.

This is the "copilot" model that's become the default framing for AI in knowledge work. It's genuinely powerful, but treating it as the whole story misses important possibilities and risks.

Analyze applies when the task itself is well-defined—scan for fraud, flag anomalies, predict equipment failures—but humans need to interpret and act on the results. AI does substantial work independently—processing data, identifying patterns, generating insights—and surfaces conclusions for human review. Think of fraud detection systems that flag suspicious transactions for human investigation, or predictive maintenance systems that alert a plant manager to equipment likely to fail.

In Analyze mode, the human isn't leading the analytical work—they're reviewing the AI's work. This requires different skills: knowing what questions to ask, understanding how the AI might be wrong, maintaining appropriate skepticism about AI-generated conclusions. The task has low ambiguity; the implications of the analysis may not.

Automate combines low task ambiguity with high AI control. The task is well-defined, the right answer is knowable, and AI handles it end-to-end without human involvement in individual decisions. Automated email sorting. Dynamic pricing systems. In warehouses, AI systems now orchestrate picking routes, manage inventory placement, and coordinate robot fleets—all without human involvement in moment-to-moment decisions.

Automate mode offers enormous efficiency gains—tasks that would be impossible to staff with humans become viable. But it requires high confidence that the AI will perform reliably, because errors may accumulate before anyone notices. The low ambiguity is what makes automation safe; if the task were genuinely ambiguous, high AI control would be dangerous.

Agent is the newest and most dynamic quadrant. Here, task ambiguity is high—the path to the goal isn't predetermined—but AI has the control to figure it out. You tell an agent to research competitors and prepare a briefing; it decides which sources to consult, what information to extract, how to synthesize findings. The AI isn't just assisting or analyzing—it's acting, navigating ambiguity autonomously toward goals you've set.

Agent mode is where AI starts to feel less like a tool and more like a worker. The governance challenge is specifying goals clearly enough that autonomous action produces good outcomes, while maintaining enough oversight to catch problems. You're trusting AI to navigate ambiguity, which means accepting that its path may differ from what you'd have chosen.

The Four A's framework isn't just descriptive, it's also a decision tool. Different situations call for different modes, and deploying the wrong mode creates unnecessary risk or leaves value on the table.

High task ambiguity with high stakes calls for Augment. When decisions matter enormously and the situation is novel or complex, you want humans firmly in control with AI supporting their judgment.

Low task ambiguity with high volume calls for Automate. When you're doing the same thing thousands of times and the right answer is well-defined, human involvement in each instance is waste.

Defined analytical tasks with human accountability call for Analyze. When AI can process information that humans can't, but humans need to own the conclusions and next steps, Analyze mode gives you both.

Ambiguous paths toward clear goals call for Agent. When the objective is clear but the route isn't, agents can figure out the how while humans specify the what.

The framework also highlights governance requirements. Augment mode needs relatively light governance—the human is right there, making judgments throughout. Automate and Autonomous modes need heavy governance—errors scale. Analyze and Agent modes fall in between, requiring thoughtful oversight structures that match the risk profile

Autonomous AI

The Four A's framework maps collaboration between humans and AI. But there's a fifth mode that sits outside the framework entirely: **Autonomous**—AI operating independently at scale, with minimal or no human involvement in individual decisions.

This isn't another collaboration mode; it's a different category. Self-driving vehicles navigating city streets. Algorithmic trading systems executing thousands of transactions per second. Dark factories where robots manufacture goods around the clock without human workers. AI systems managing power grids, optimizing logistics networks, or coordinating drone fleets.

Humans design the system, set high-level objectives, and monitor aggregate performance, but they're not collaborating with AI on individual tasks. AI largely operates on its own.

Autonomous AI raises distinct challenges: accountability when things go wrong, transparency into AI decision-making, the ability to intervene when needed. For most leaders today, autonomous AI isn't yet a priority, but if you're in logistics, manufacturing, or transportation, these challenges are immediate.

AI Teammates

Here's a sentence that would have sounded absurd five years ago: "Steve handles our financial modeling. Steve is an AI. Steve attends our standups and has deliverables."

This isn't science fiction. Companies are already experimenting with AI agents that have persistent identities, defined responsibilities, and integration into organizational workflows. They have names. They appear in Slack channels. They're assigned to projects. They produce work that other team members depend on.

AI teammates represent what mature agent deployment looks like when it gains persistent identity. Rather than invoking AI for discrete tasks, you're integrating AI into your organization as an ongoing participant—something closer to a team member than a tool.

The potential is significant. AI teammates can handle work that's too voluminous for humans, too tedious to sustain attention, or too continuous for shift-based staffing. They free human teammates to focus on work that requires human capabilities.

The caution is equally important. AI teammates don't have judgment in the way humans do. They don't understand context they haven't been given. They can fail in ways that look like success until someone notices the accumulating errors.

TRY THIS

Pick one role on your team that involves substantial routine work. Ask: "If I wanted an AI teammate to handle this, what would it need to do? What oversight would be required? What could go wrong?"

Some practical questions: What happens when an AI teammate makes a mistake? Who's accountable? How do you maintain quality when the work is continuous and voluminous? How do you prevent other team members from over-trusting AI outputs?

For now, many organizations are experimenting cautiously: giving AI teammates limited responsibilities, maintaining close human oversight, learning what works before expanding scope.

This seems wise.

When Agents Work Together

So far, we've considered AI agents as individual actors—a single agent researching competitors, analyzing documents, or managing a workflow. But the frontier is rapidly moving toward something more complex: multiple agents working together, each with specialized capabilities, coordinating to accomplish goals that no single agent could handle alone.

This is already happening in software development, where AI coding agents now work alongside testing agents, documentation agents, and code review agents—each handling a piece of a larger workflow. It's emerging in customer service, where a front-line AI agent handles initial inquiries, escalates complex issues to a specialist agent, and hands off to a fulfillment agent when action is required. It's appearing in research workflows, where one AI agent gathers information, another synthesizes it, and a third critiques the synthesis for gaps and errors.

The pattern resembles how human organizations work: specialization, handoffs, coordination. But agents can work in parallel (simultaneously rather than sequentially) to operate at speeds and scales that human teams cannot match—and they introduce coordination challenges that are genuinely new.

Orchestrating Agents

When multiple agents work together, someone—or something—needs to coordinate them. This orchestration layer decides which agent handles which task, manages handoffs, resolves conflicts when agents produce contradictory outputs, and ensures the overall goal stays on track.

Today, humans typically play this orchestration role. You might use one agent to research a topic, manually review its output, then hand selected pieces to another agent for synthesis. The coordination happens in your head and through your actions.

But orchestration is increasingly being automated too. AI systems can now coordinate other AI systems—deciding when to invoke which agent, routing information between them, and synthesizing their outputs. This creates layers: agents doing work, orchestrator agents coordinating that work, and humans overseeing the orchestrators.

For leaders, this raises a critical question: Where in this stack does human judgment need to sit? The answer isn't obvious, and it will vary by situation. High-stakes decisions probably need human oversight closer to the action. Routine workflows might be safe to delegate entirely to agent orchestration. Getting this right is a new management skill—one that didn't exist five years ago.

Within this new environment, several multi-agent patterns are emerging.

Swarm. Multiple agents tackle the same task simultaneously and independently, with the best result selected or all results aggregated. Think of it like assigning the same brief to several teams and choosing the strongest proposal—or combining the best elements from each. This pattern trades efficiency for quality, and it's particularly useful when the "right" answer isn't obvious upfront. Creative tasks, strategy generation, and complex problem-solving benefit from this parallel exploration.

Roundtable. Multiple agents analyze the same problem from different angles, then their outputs are synthesized—often by another agent. This deliberately introduces productive disagreement. One agent might argue for a course of action while another critiques it. The goal isn't consensus but comprehensive analysis.

Auditor. One set of agents do the work; another set of agents review it. This simple pattern catches errors, reduces hallucinations, and adds a layer of quality control. It's particularly valuable for high-stakes outputs where mistakes are costly.

Multi-agent systems aren't something most organizations need to build from scratch—at least not yet. The platforms and tools are maturing rapidly, and in many cases you'll adopt multi-agent capabilities through the products you already use rather than assembling them yourself.

Understanding that AI agents can work in teams, not just as individuals, changes how you think about what's possible. Tasks that seem too complex for "an AI" might be tractable for a coordinated system of specialized agents. Workflows you assumed required human coordination because of their complexity may be candidates for agent orchestration.

It changes how you evaluate AI tools. When vendors describe their AI capabilities, asking "how do your agents work together?" reveals whether they're thinking about individual tasks or coordinated workflows.

It raises governance questions you'll need to answer. When something goes wrong in a multi-agent system, accountability is harder to trace. Oversight mechanisms need to account for agent interactions, not just individual agent behavior.

Jobs Transformed, Jobs Eliminated

The headlines focus on job losses: AI will eliminate millions of positions. The counternarrative focuses on job creation: new technologies always create more jobs than they destroy.

The truth is more nuanced. History suggests that major technological transitions eliminate some jobs, transform others, and create new ones. The industrial revolution eliminated countless artisan positions while creating factory jobs. The computer revolution eliminated typing pools while creating entirely new industries. These transitions were genuinely disruptive, causing real hardship, while also generating long-term prosperity.

AI will likely follow this pattern. Some jobs will be eliminated— particularly those consisting primarily of tasks AI can now perform more cheaply. Some jobs will be transformed—the tasks change, but the role persists in new form. Some jobs will be created—new roles serving needs we can't fully anticipate.

Consider what this looks like in practice. Imagine a warehouse operations supervisor who spends most of her day on scheduling, route optimization, and inventory tracking. AI systems begin handling all of this — and handling it better than she ever could. Her initial fear is understandable: her job is disappearing. But look at what emerges. She now focuses on exception handling when AI systems encounter situations they can't resolve, on training workers who interact with AI-driven systems, on identifying process improvements that AI can't see because they require understanding human factors, and on managing the interface between automated systems and the humans who work alongside them.

The pattern is instructive: routine tasks migrated to AI, and her role shifted toward judgment, human connection, and contextual problem-solving. Unfortunately, not every transition will be this smooth—some roles will genuinely disappear. But "transformation" isn't just a euphemism. It's often what happens.

For leaders, this means holding multiple realities simultaneously.

First, the human reality. The people in your organization aren't abstractions. They're individuals who built their careers developing capabilities that may now be less valuable. Leading through transformation means honoring their contributions while being honest about how the world is changing.

Second, the organizational reality. Your organization needs capabilities that serve customers and sustain competitive advantage. Clinging to staffing models that no longer make sense isn't kindness—it's a slow path to organizational failure.

Third, the adaptive reality. The transitions that go well are those where leaders invest in helping people develop new capabilities, create pathways to new roles, and treat affected individuals with dignity.

The executives who navigate this well will be those who resist both the "everything will be fine" complacency and the "nothing can be done" fatalism. This is a "yes, and" moment: yes, AI will displace some work, and we can shape how that displacement happens, and here's how we're going to do it.

What Are You Really Contributing?

The organizational questions about jobs eventually become personal. If AI can do much of what you do, what's left for you to contribute?

AI accelerates the exploration of possibilities. The architect still decides what gets built. Their contribution has shifted, but it hasn't disappeared—if anything, it's been clarified. The value lies in judgment, not execution.

Consider the surgeon. Surgeons have always used tools—scalpels, laparoscopes, imaging systems. Now AI assists with diagnosis, surgical planning, even robotic precision during procedures. Their value is judgment under pressure—the ability to adapt when things go wrong, to integrate information in real-time. It's also the relationship with patients, the accountability for outcomes, the human presence in moments of vulnerability.

The pattern holds across roles. The analyst who can interpret what information means remains essential. The manager who can motivate, develop people, and exercise judgment remains essential. The specialist who can apply judgment, handle edge cases, and take accountability remains essential.

TRY THIS

List your five most important professional capabilities. For each one, ask an AI how well current systems perform it. Be honest about the answers. Where AI falls short, you've identified your distinctive contribution.

The question "what are you really contributing?" isn't an attack. It's an invitation to clarity. When you strip away the tasks that AI can now perform, what's left? That remainder—judgment, relationships, accountability, creativity, wisdom—is your actual value proposition. AI doesn't threaten it. AI reveals it.

Building AI-Ready Organizations

The individual question—what are you contributing?—scales to an organizational question: what kind of workforce do you need?

This means investing in AI literacy across the organization. Not turning everyone into engineers, but helping everyone understand what AI can and can't do, how to use AI tools effectively, and how to maintain appropriate judgment about AI outputs.

It means redesigning work to take advantage of human-AI collaboration. Not just adding AI tools to existing processes, but rethinking how work flows to optimize for what humans and AI each do best.

It means creating psychological safety around AI adoption. People who fear that learning AI will make them obsolete won't engage authentically. Leaders need to create environments where experimentation is safe, where questions are welcome.

And it means being honest about the transition. AI will change roles. Some positions will evolve; some will be eliminated. Pretending otherwise destroys trust. Acknowledging the reality— while demonstrating commitment to managing the transition responsibly—builds credibility.

What Remains

Let's close with a reframe. Much of the discourse around AI and work focuses on what AI takes away—the jobs eliminated, the tasks automated. This framing is understandable but incomplete.

There's another way to see it: AI clarifies what's distinctively human.

When AI can analyze, humans who synthesize become more valuable. When AI can generate, humans who curate become more valuable. When AI can process, humans who exercise judgment become more valuable. When AI can simulate relationships, genuine human connection becomes more precious.

The human premium isn't disappearing. It's becoming more visible. For decades, human capabilities were bundled with tasks that happened to require humans because no alternative existed. Now the bundle is coming apart. The tasks that were never really about human capabilities are migrating to AI. What remains is what was always most important.

This is disruptive. Transitions are hard. But for those who can identify and develop their distinctively human contributions, the AI era offers the possibility of work that's more meaningful, not less—work that consists of what humans do best.

The companies that thrive will be those that understand this—that see AI not as a replacement for humans but as a clarifier of human value.

Part II examined how AI changes competition, business models, and the nature of work. These are the strategic questions—the "what" and "why" of AI transformation. In Part III, we turn to leadership questions—the "how" of making it happen responsibly. Governance, investment, and execution await.

Part III

LEADING RESPONSIBLY

Vision gets the applause. Governance, investment, and execution get the results.

7

GOVERNANCE, ETHICS, AND RISK

We start with governance because everything else depends on it. Investment decisions require frameworks for evaluating risk. Organizational change requires clarity about boundaries. Execution requires knowing what's permissible and what isn't. Get governance wrong, and even brilliant strategy leads to disaster.

Many executives hear "governance" and think "bureaucracy"—red tape that slows things down, compliance requirements that constrain innovation, legal overhead that adds cost without value. This is a misunderstanding, and it's a dangerous one.

I understand the frustration firsthand. I've had promising AI initiatives blocked by security restrictions and regulatory requirements that felt like they existed to prevent progress rather than enable it. The experience of watching a good idea die in a compliance review is genuinely maddening. But I've also seen what happens when organizations move fast without governance; the cleanup is often more expensive than the delay.

Governance is what lets you move fast without breaking things that matter.

Consider the alternative. A financial services firm deploys an AI system that inadvertently discriminates against certain applicants. A healthcare company's diagnostic AI makes recommendations based on biased training data. A retailer's pricing algorithm engages in patterns that regulators consider predatory. A professional services firm floods the market with AI-generated content that damages its brand. In each case, initial deployment is fast—and the eventual cleanup is slow, expensive, and reputation-destroying.

Trust carries an economic dimension that's easy to overlook: low-trust environments are expensive. When people don't trust results, they double-check everything. When teams don't trust each other's work, they duplicate effort. When customers don't trust companies, they demand guarantees or go elsewhere.

This is the trust tax. It applies everywhere, but it compounds in AI-driven environments where outputs are harder to trace. Organizations that invest in AI governance—transparency, accuracy, security—will pay a lower trust tax and move faster as a result.

Delegating AI Governance

Here's the uncomfortable truth that many executives don't want to hear: AI governance is your job. Not your CTO's job. Not your chief data officer's job. Not your legal team's job. Yours.

This doesn't mean you personally configure every AI system or review every deployment. It means you're accountable for the framework, the standards, and the culture that shapes how your organization uses AI. You can delegate tasks, but you can't delegate responsibility.

Why can't this be delegated? Three reasons.

First, AI governance involves value judgments that only leadership can make. What level of risk is acceptable? Which trade-offs between efficiency and safety are appropriate? How much should we invest in governance relative to capability? These aren't technical questions—they're strategic choices that reflect organizational values. An engineer can tell you whether an AI system achieves 95% or 99% accuracy. Only leadership can decide whether 95% is good enough for this particular application, given the consequences of the other 5%.

Second, AI governance requires organizational authority. Effective governance means sometimes saying no to projects that would be profitable, sometimes slowing down initiatives that could move faster, sometimes enforcing standards that departments find inconvenient. This requires the authority to make decisions that not everyone will like—and to make them stick. Technical teams don't have that authority. Legal teams don't have that authority. Only executives do.

Third, AI governance sets the tone for organizational culture. How leadership talks about AI ethics, how seriously they take governance reviews, how they respond when problems arise—all of this signals what really matters. If executives treat governance as a checkbox exercise, the organization will too. If executives visibly engage with difficult ethical questions, take governance seriously, and hold people accountable, that becomes the culture. Tone at the top isn't a cliché; it's how organizations work.

The executives who get this right aren't the ones who understand AI best technically. They're the ones who recognize that AI governance is leadership work, not technical work, and who own it accordingly.

Building a Governance Framework

What does an AI governance framework look like? The specifics vary by organization, but effective frameworks share common elements.

Clear principles that articulate what your organization believes about AI use. Not vague platitudes—"we believe in responsible AI"—but specific commitments that guide decisions. What do you believe about transparency to customers? About human oversight of automated decisions? About the trade-off between personalization and privacy? These principles should be concrete enough that someone facing a difficult decision can use them.

Defined accountability for AI decisions. When an AI system makes a consequential choice, who's responsible? This question is harder than it sounds. Is it the team that built the model? The team that deployed it? The team that uses it? The answer matters enormously when something goes wrong. Effective governance makes accountability explicit before problems arise, not after.

Risk classification that distinguishes high-stakes from low-stakes applications. Not every AI deployment requires the same level of scrutiny. An AI that recommends internal meeting times needs less governance than one that approves loans or diagnoses diseases. A good framework categorizes applications by risk level and calibrates oversight accordingly. This prevents both recklessness (treating everything as low-risk) and paralysis (treating everything as high-risk).

Review processes that are proportionate to risk. High-risk applications might require formal review boards, external audits, and ongoing monitoring. Low-risk applications might need only lightweight documentation and periodic spot-checks. The goal is appropriate scrutiny—enough to catch problems, not so much that innovation grinds to a halt.

Mechanisms for escalation when edge cases arise. No framework anticipates every situation. Effective governance includes clear paths for raising difficult questions—and a culture that encourages people to use them. The alternative is that edge cases get resolved at whatever level first encounters them, which often means they're resolved badly.

Regular review and adaptation as circumstances change. AI capabilities evolve rapidly. Regulatory requirements shift—what's voluntary today may be mandatory tomorrow. Build in mechanisms for periodic review and update on a regularly-scheduled cadence. Design your governance to anticipate tightening requirements, not just meet current ones.

The goal isn't to create bureaucratic overhead. It's to create clarity that enables faster, better decisions. When people know what's expected, what's allowed, and who decides edge cases, they can move confidently. When they don't, every decision becomes fraught.

AI Risk Taxonomy

Effective governance requires understanding what can go wrong. AI risks cluster into several categories, each requiring different mitigation strategies.

Performance risks are the most obvious: the AI simply doesn't work as intended. It makes inaccurate predictions, generates flawed content, or fails to handle edge cases. These risks are relatively well-understood and manageable through testing, monitoring, and human oversight. The challenge is maintaining vigilance—performance can degrade over time as conditions change, and failures may be subtle rather than obvious.

Bias and fairness risks arise when AI systems treat different groups differently in ways that are unjust or illegal. An AI trained on historical data may perpetuate historical biases. A system that works well on average may work poorly for minority populations underrepresented in training data. These risks are harder to detect than simple performance failures—the system may appear to work well while quietly discriminating.

Privacy risks emerge from AI's hunger for data. Training effective AI often requires vast amounts of information, some of which may be personal or sensitive. Models can inadvertently memorize and later reveal private information. Predictions themselves may reveal more than customers intended to share. As AI systems become more capable, privacy risks intensify.

Security risks include vulnerabilities specific to AI systems. Adversarial attacks can manipulate AI behavior in ways that wouldn't affect traditional software. Training data can be poisoned. Models can be extracted or reverse-engineered. And as AI moves from cloud-based chatbots to autonomous agents and locally installed applications, the attack surface expands further — every tool that can take action on behalf of your organization is a tool that can be manipulated into taking the wrong action.

Transparency and explainability risks arise when stakeholders (e.g., customers, regulators, employees) don't understand how AI decisions are made. Some AI systems are inherently opaque; their decisions emerge from complex patterns that resist simple explanation. This opacity creates problems when decisions need to be justified, appealed, or audited.

Dependency and concentration risks emerge as organizations become reliant on AI systems or AI vendors. What happens if a critical AI system fails? What if your AI provider changes terms, raises prices, or goes out of business? What skills atrophy as humans stop performing tasks that AI now handles? These risks are often invisible until they materialize.

Reputational risks connect all of the above to stakeholder perception. An AI failure that might be technically minor can become a major reputational event if it involves bias, privacy violations, or harm to vulnerable populations. The court of public opinion operates by different rules than technical performance metrics.

A good governance framework addresses each category explicitly. For each type of risk, you need to know: What are we doing to prevent it? How would we detect it if it occurred? How would we respond? Who's responsible?

TRY THIS

Pick an AI application your organization uses or is considering. Walk through each risk category and ask questions. How exposed are we? What's our mitigation strategy? Who would know if something went wrong?

Bias and Fairness

Among AI risks, bias and fairness deserve particular attention— not because they're more likely than other risks, but because they're more complicated and more consequential when they occur.

The fundamental challenge is this: AI systems learn patterns from historical data. If that data reflects historical biases (how can it not?) the AI will learn those biases. A job screening AI application trained on hiring data may learn that certain demographics were hired at higher rates and may perpetuate that pattern. An AI trained on lending data will learn which borrowers defaulted, without accounting for whether lending practices themselves contributed to the default rates.

This creates a genuine dilemma. The AI isn't "wrong" in a statistical sense—it's accurately predicting outcomes based on historical patterns. But those predictions perpetuate injustices embedded in the historical data. Statistical accuracy and fairness can point in different directions.

There's no purely technical solution to this problem. You can measure AI systems for different kinds of bias. You can test for disparate impact. You can adjust training data or model outputs to reduce measured bias. But every intervention involves trade-offs, and those trade-offs are value judgments. Should an AI system achieve equal accuracy across groups, even if that means lower overall accuracy? Should it achieve equal selection rates, even if that means different accuracy rates? There's no objectively correct answer—only choices that reflect what you value.

This is why AI fairness can't be delegated to technical teams. They can implement whatever fairness criteria you specify. But choosing those criteria is a leadership decision.

Practical guidance for thinking about AI fairness:

Start with the question: "What would fair look like in this context?" The answer varies by domain. Fair hiring might mean something different from fair lending might mean something different from fair healthcare. Don't assume a one-size-fits-all definition.

Consider who's affected and how. AI fairness isn't abstract— it affects real people. Which groups might be disadvantaged by this system? What are the consequences of unfair treatment? The answers should inform both how much scrutiny the system gets and what fairness standards apply.

Be honest about trade-offs. There's often no option that maximizes both accuracy and fairness, or that satisfies all reasonable definitions of fairness simultaneously. Leaders who pretend these trade-offs don't exist aren't leading—they're avoiding.

81

Document your reasoning. When fairness trade-offs arise, document what you considered, what you decided, and why. This isn't just legal protection—it forces clarity of thought and provides a basis for future review.

Expect evolution. Our understanding of AI fairness is developing rapidly. What seems like a reasonable approach today may be recognized as inadequate tomorrow. Build in mechanisms to revisit fairness decisions as understanding advances.

Governing Excess

Most conversations about AI governance focus on high-stakes decisions: who gets a loan, who gets hired, who receives medical treatment. These are important. But there's a different governance challenge that receives less attention and may cause more aggregate harm: the flood of low-quality AI-generated content.

Organizations can become slop factories without anyone intending it. The dynamics are insidious. AI makes content production cheap, so more content gets produced. No individual piece seems obviously bad—it's grammatically correct, professionally formatted, superficially plausible. But cumulatively, the content adds no value while consuming audience attention and eroding brand trust. The damage accumulates slowly, then suddenly.

This is a governance problem because it requires organizational standards that individuals can't set alone. A marketing manager trying to meet content quotas will use AI to produce more content—that's rational individual behavior. But if everyone does this without coordination, the organization floods its channels with mediocrity. Only governance—organizational standards enforced with authority—can prevent the race to the bottom.

What does governance for content quality look like?

Quality standards that are explicit and enforced. "Good enough" is too vague. What specifically makes content worth publishing? What review processes catch content that doesn't meet standards? Who has authority to reject content that technically meets metrics but adds no real value?

Metrics that measure value, not just volume. If your metrics reward content production, you'll get content production. If they reward engagement, impact, or customer value, you'll get something better. Governance means ensuring that metrics align with actual organizational goals, not just easy-to-count outputs.

Permission to produce less. Sometimes the right answer is not to publish. Governance should make it acceptable—even praiseworthy—to decide that something isn't worth producing, even if producing it would be easy. This is harder than it sounds; the pressure to do something with available capacity is real.

Human judgment at key points. AI can draft content, but humans should decide what's worth saying and whether the draft says it well. This doesn't mean humans review every word—that defeats the efficiency gains. But it means humans are in the loop at decision points that matter.

The slop problem is a "yes, and" challenge. Yes, AI makes content creation cheap—and that requires new disciplines to prevent abundance from becoming pollution. Organizations that get this right will maintain brand integrity while their competitors drown in their own mediocrity.

Transparency & Explainable AI

When an AI system makes a decision that affects someone— denying a loan, flagging a transaction, recommending a diagnosis—they deserve to understand why. Regulators increasingly require it. Customers expect it. Ethics demand it.

But AI systems don't always make this easy. Some of the most powerful models are also the most opaque. A deep neural network might achieve remarkable accuracy while offering little insight into how it reaches conclusions. The patterns it learns may be too complex for humans to interpret, or may not correspond to concepts humans recognize.

This creates tension between capability and transparency. The most accurate model may be the hardest to explain. The most explainable model may perform worse. How you navigate this tension depends on context.

For high-stakes decisions affecting individuals—credit, employment, healthcare—explanation is non-negotiable. Regulatory requirements aside, you simply can't defend a consequential decision with "the algorithm said so." This may mean accepting some accuracy trade-off in favor of interpretable models, or building explanation layers around opaque models, or ensuring human oversight that can provide explanation even when the AI can't.

For lower-stakes applications—content recommendations, operational optimizations, internal productivity tools—less explanation may be acceptable. The person affected by a product recommendation doesn't necessarily need to know why. The cost of a wrong recommendation is low enough that accountability can be aggregate rather than individual.

Governance should specify explanation requirements by application type. For each AI deployment, ask: Who might need to understand this decision? What would adequate explanation look like? Can our system provide that explanation? If not, what changes are needed—to the system, to the oversight structure, or to the deployment decision itself?

Intellectual Property

There's a category of AI risk that doesn't fit neatly into bias, privacy, or security—but that's generating real legal exposure right now: intellectual property.

The questions are multiplying faster than the answers. If an AI system was trained on copyrighted material—and most large language models were—does using that system expose you to infringement claims? If AI generates content for your organization, who owns it? If AI generates something substantially similar to an existing copyrighted work, who's liable?

84

Courts and legislatures are working through these questions, but the legal landscape remains unsettled. Several high-profile lawsuits are challenging the legality of training AI models on copyrighted works. The US Copyright Office has issued guidance indicating that AI-generated content without meaningful human authorship may not be copyrightable—which raises questions about whether your AI-assisted deliverables are protectable.

You don't need to resolve every open legal question—no one can, because the law is still catching up. But you do need governance that manages the uncertainty. That means understanding what AI tools your organization uses and what their training data and licensing terms look like. It means having clear policies about ownership and attribution for AI-assisted work, especially for client-facing deliverables. It means assessing where your organization has the most IP exposure—usually content generation, code production, and creative work—and applying appropriate review processes.

If your proprietary content—documents, designs, code, and customer data—is being used to train AI systems you interact with, that's an IP governance question too. And it means staying close to the legal developments, because the answers that emerge over the next few years will reshape what's permissible and what's risky.

These policies will need to evolve as the legal landscape settles— what matters now is having a framework for thinking about the questions, not having all the answers.

Governing Agents

Most AI governance frameworks were designed for AI that analyzes and recommends. Agents are different. They don't just suggest—they act. They make decisions, use tools, and produce consequences in the real world. This requires governance mechanisms that go beyond reviewing AI outputs to controlling AI behavior.

Consider these dynamics when planning for agent governance:

Establish boundaries. With conversational AI, governance focuses on what the AI produces—reviewing outputs for accuracy, bias, or appropriateness. With agents, governance must also define what the AI is permitted to do. What systems can it access? What actions can it take? What's explicitly off-limits?

Think of this like defining a job description and authority limits for a new employee—except you need to be far more explicit. Humans understand implicit boundaries; agents follow their explicit permissions. An agent told to "optimize customer response time" without constraints might take actions that technically achieve the goal but violate norms you assumed were obvious.

Clearly established boundaries should include: what data the agent can access, what external systems it can interact with, what types of actions it can take, what spending or commitment authority it has, and what conditions should trigger escalation to humans.

Audit trails for autonomous decisions. When an agent takes a sequence of actions to accomplish a goal, you need to be able to reconstruct what it did and why. This isn't just about debugging when things go wrong—it's about accountability, compliance, and organizational learning.

Effective agent audit trails capture: the goal or instruction the agent received, the steps it took and in what sequence, the reasoning behind key decisions (to the extent the agent can articulate it), what information it accessed, what external actions it took, and what the outcomes were.

This logging needs to happen automatically, not as an afterthought. Build it into agent deployment from the start.

Intervention mechanisms. What happens when an agent is heading in the wrong direction? You need the ability to pause, redirect, or stop agent activity—ideally before consequences become irreversible.

This means designing for interruption. Agents should have clear checkpoints where human review can occur. High-stakes or irreversible actions should require explicit confirmation. There should be a way to halt an agent mid-task if problems emerge. And rollback mechanisms should exist for actions that can be undone.

The goal isn't to review every agent action—that would defeat the purpose of automation. It's to ensure that when intervention is needed, it's possible.

Multi-agent accountability. When multiple agents work together, accountability becomes distributed. If an outcome is wrong, which agent is responsible? The one that gathered the information? The one that synthesized it? The orchestrator that coordinated them? The human who set the goal?

There's no universal answer, but you need a framework for your organization. Generally, accountability should rest with whoever had the authority and information to prevent the problem. In practice, this means humans must remain accountable for outcomes even when agents execute the work—which has implications for how much oversight humans need to maintain.

As agent capabilities grow, this tension will intensify. Organizations that think carefully about accountability now will be better positioned than those who wait for a crisis to force the question.

Making Governance Operational

Principles and frameworks matter, but governance only works if it's operational—if it shapes behavior throughout the organization. How do you make that happen?

Embed governance in workflows, not around them. Governance that requires people to step outside their normal work processes won't be followed consistently. Governance that's built into how work gets done becomes automatic. If deployment requires governance review, embed that review within the deployment process, not as an extra step people might skip.

Make governance proportionate to risk. Excessive governance for low-risk applications creates two problems: it wastes resources, and it trains people to view governance as

87

obstacle rather than enabler. If everything requires full review, nothing feels important. Reserve heavy governance for applications that genuinely warrant it.

Create feedback loops. Governance isn't set-and-forget. You need mechanisms to learn whether governance processes are working, whether they're catching problems, whether they're creating unnecessary friction. Build in ways to measure and improve.

Invest in governance capability. Governance requires skills: understanding AI risks, evaluating bias, assessing trade-offs. If you expect business leaders to make governance decisions, give them the knowledge and tools to make good ones. Don't just issue policies; build capacity to implement them.

Lead visibly. When executives engage with governance questions—asking hard questions in reviews, taking ethical concerns seriously, making difficult calls when trade-offs arise—it signals that governance matters. When they delegate everything to legal or compliance, it signals the opposite.

TRY THIS

In a protected AI environment, ask for a review of your governance documents. How do they align with industry standards? Where are the gaps and vulnerabilities?

Governance Readiness

Let me offer a simple diagnostic. For each AI application your organization uses or is considering, can you answer these questions?

What decisions does this AI make, and what are the stakes? Who's accountable when those decisions go wrong? What could fail, and how would we know? What oversight is in place, and is it proportionate to the risk? How do we explain AI decisions to those affected? What are our standards for fairness, and how do we measure compliance? What's our process for reviewing and updating governance as conditions change?

If you can answer these questions clearly for each significant AI application, your governance is in reasonable shape. If you can't, you've identified work to do.

Perfect governance doesn't exist. The goal is thoughtful engagement with the questions, honest assessment of gaps, and continuous improvement.

The Regulatory Landscape

Everything we've discussed so far is governance you choose, but a growing portion of AI governance is mandatory.

The European Union's AI Act represents the most comprehensive AI regulation to date. It classifies AI systems by risk level—from minimal risk, which faces few requirements, to unacceptable risk, which is banned outright—and imposes obligations that escalate accordingly. High-risk applications in areas like employment, credit, education, and law enforcement face requirements for transparency, human oversight, data governance, and conformity assessments. Organizations that deploy AI systems affecting people in the EU must comply regardless of where they're headquartered. If you operate internationally, this isn't optional reading.

In the United States, the regulatory approach has been more fragmented—a patchwork of executive orders, sector-specific guidance, and state-level legislation rather than a single comprehensive framework.

Federal agencies are increasingly applying existing authorities to AI: the FTC is scrutinizing AI-related consumer harms, the EEOC is examining AI in hiring decisions, and financial regulators are tightening expectations around AI-driven lending and trading.

Meanwhile, states are moving independently. The result is a complex and shifting landscape where compliance requirements depend heavily on where you operate, what sector you're in, and what your AI systems do.

Three principles can help you navigate this complexity without getting lost in the details.

Build ahead of requirements, not behind them. The governance you build because it's wise often turns out to be governance that satisfies regulators. That's not a coincidence; good governance and regulatory compliance point in the same direction.

Focus on principles, not jurisdictions. You can't track every regulatory development in every market. But you can build governance around principles—transparency, accountability, fairness, human oversight, proportionate risk management—that satisfy the intent of regulation across jurisdictions.

Make regulatory awareness a leadership function, not just a legal one. Your legal team needs to track the specifics. But the strategic implications of AI regulation—which markets to enter, which applications to deploy, how to position your organization—are leadership decisions. An executive who doesn't understand the regulatory landscape is making strategic choices without seeing the full board.

One final note: regulation can be a competitive variable, not just a compliance burden. Organizations that invest early in governance and compliance build trust with customers, partners, and regulators.

Governance as Enabler

Let's close where we started. Governance isn't bureaucracy. It's not the thing that slows you down. Done right, it's the thing that lets you move fast sustainably.

The companies that will lead in the AI era aren't the ones that move fastest with least oversight. They're the ones that build governance frameworks enabling them to move confidently— knowing they can identify and mitigate risks, knowing they can explain and defend their decisions, knowing they can catch problems before they become crises.

This is the "well" in "leading AI well." Not just competence, but conscience. Not just capability, but responsibility. Not just moving fast and breaking things with glee, but moving thoughtfully.

Good governance is what makes that possible. It's not overhead; it's infrastructure for sustainable AI leadership.

Now that we have a framework for governing AI responsibly, we need to address the other constraint every leader faces: resources. AI requires investment—but how do you make investment decisions when outcomes are uncertain and the landscape keeps shifting?

8

INVESTMENTS AND RETURNS

AI can deliver dramatic performance improvements, but such returns don't come for free. At some point, you have to spend money—on tools, on talent, on infrastructure, on experiments that might fail. And spending money means making investment decisions.

This is where many executives get stuck. The AI landscape feels too uncertain for traditional capital allocation. The technology changes too fast. The outcomes are too unpredictable. The hype is too loud to hear the signal. So they wait for clarity—and while they wait, competitors are building capabilities, developing intuitions, and learning lessons that can't be acquired any other way.

Executives have to make AI investment decisions under uncertainty—not despite the uncertainty, but in ways that account for it. We'll examine how to think about ROI when outcomes are genuinely unpredictable, how the economics of AI are shifting faster than most executives realize, and how to evaluate AI investments in ways that survive skepticism and budget pressure.

But first, we need to confront the investment question that most aren't asking: What's the cost of doing nothing?

The Invisible Cost of Inaction

Every investment decision is really two decisions: the cost of acting and the cost of not acting. We're trained to scrutinize the first. We build spreadsheets, model scenarios, calculate payback periods. But the second cost—the cost of inaction—is invisible. It doesn't show up in any budget. No one submits an expense report for capabilities never built.

This invisibility biases every AI investment discussion. The executive who proposes an AI initiative must defend projected costs and uncertain returns. The executive who proposes doing nothing defends... nothing. Inaction requires no business case. It faces no scrutiny. It just happens, quarter after quarter, while the organization falls further behind.

But the cost of inaction is real, even when it's invisible. It shows up in the changes you never make because you lacked the AI fluency to see they were possible. It shows up in the capabilities you never build because you were waiting for certainty that never comes. It shows up in the intuitions you never develop because intuition requires experience, and experience requires action. It shows up in the mistakes you never learn from because learning requires trying, and trying requires investment.

Consider what your competitors are learning right now. Every organization experimenting with AI is generating knowledge—about what works in their context, what doesn't, how to implement effectively, where the pitfalls lie. This knowledge compounds. The organization that starts building AI capabilities today will be dramatically better at building AI capabilities two years from now. The organization that waits will face the same learning curve later, except now they're learning while competing against rivals who've already climbed it.

The cost of inaction also compounds in talent. The best AI-fluent people want to work on interesting problems with committed organizations. They can tell when a company is serious about AI and when it's going through the motions. Every year you delay building genuine AI capability is a year the talent market gets more competitive and your organization gets less attractive to the people who could help you catch up.

And there's a subtler cost: the cultural cost of hesitation. Organizations that treat AI as something to be studied rather than something to be used develop cultures of caution. People learn that proposing AI initiatives is risky—lots of scrutiny, uncertain outcomes, career exposure. So they stop proposing. The organization's collective imagination about what's possible shrinks. By the time leadership decides to act, they've inadvertently trained their people to wait for permission.

None of this appears in a financial model. But it's the real cost of the "let's wait and see" approach that feels so prudent in the moment.

Shifting Economics of AI

If you've been following AI investment news, you've probably seen the headlines about massive datacenter builds—billions of dollars in infrastructure, eye-watering energy consumption, supply chains straining to produce enough chips. These stories create an impression that AI is enormously expensive, accessible only to the largest companies with the deepest pockets.

That impression is increasingly wrong, and misunderstanding the economics is leading many organizations to delay investments they should be making now.

Here's what the datacenter headlines miss: the cost of using AI— inference costs, in technical terms—has been plummeting. The expensive infrastructure investments are being made by a handful of companies building foundation models. But the cost of deploying those models, of running AI applications in your business, has dropped dramatically and continues to fall. Tasks that cost dollars two years ago cost pennies today. Capabilities that required dedicated infrastructure can now be accessed through APIs at commodity prices.

Simultaneously, powerful open-source models have become freely available. Organizations that once would have needed to license expensive proprietary systems or build from scratch can now deploy sophisticated AI capabilities using models that are free to use and modify. The barrier isn't access to technology anymore— it's knowledge of how to use it effectively.

This shift in economics changes the investment calculus fundamentally. The question isn't whether you can afford the technology—increasingly, you can. The question is whether you're investing in the organizational capabilities to use it: the talent to implement effectively, the governance to deploy responsibly, the processes to learn and iterate, the culture to experiment productively.

The executives waiting for AI to get cheaper before investing are solving the wrong problem. The technology is already cheap enough. What's expensive is catching up after your competitors have spent years learning how to use it.

Investing in Uncertainty

Traditional ROI analysis assumes you can project costs and benefits with reasonable confidence. You invest X, you expect to return Y, you calculate whether Y justifies X. This works well for mature technologies with predictable outcomes—a new manufacturing line, a fleet upgrade, an ERP implementation.

AI investments often don't fit this model. The technology is evolving too fast. The outcomes are too context-dependent. The range of possible results—from transformative success to expensive failure—is too wide. Executives trained on traditional capital allocation find themselves paralyzed: how do you approve an investment when you can't credibly project the returns?

The answer is to stop treating AI investments like capital projects and start treating them like learning investments. The goal isn't to predict outcomes with precision; it's to buy the option to learn, iterate, and scale what works.

This reframe has practical implications for how you structure AI investments.

Fail fast, not big. Instead of large initiatives with long timelines and binary success/failure outcomes, structure investments as portfolios of smaller experiments. Each experiment should be designed to generate learning quickly—weeks, not years. Most will fail to produce transformative results. That's fine. The goal is to find the ones that work before you've bet the farm on any single approach.

Iterate toward value. AI implementations rarely work perfectly on the first try. The organizations that succeed are the ones that expect iteration and build it into their investment approach. Budget for version two and version three, not just version one. Measure progress in learning velocity, not just immediate returns.

Separate experimentation from scaling. The investment profile for trying something new is different from the investment profile for scaling something proven. Experimentation budgets should be sized for learning—small enough to try many things, with clear criteria for what would justify larger investment. Scaling budgets come later, when you've validated that something works in your context.

Value optionality. In uncertain environments, options have value. An AI investment that doesn't produce immediate returns but builds capabilities you'll need later has real value—even if that value doesn't show up in a traditional ROI calculation. The organization that has experimented with AI assistants, built some internal expertise, and learned what works is better positioned than one starting from zero, even if neither has a "successful" deployment yet.

If we shift our perspective from AI as a capital project to AI as a learning investment, consider the following questions.

What are we trying to learn? Every AI investment should have a learning objective, not just a business objective. What hypothesis are we testing? What will we know afterward that we don't know now? If the initiative "succeeds," what will we have learned? If it "fails," what will we have learned? If you can't articulate the learning objective, you're not ready to invest.

How quickly will we learn it? Time is a resource. An initiative that generates learning in three months is worth more than one that takes two years, even if the eventual learning is similar. Structure investments to produce insights quickly. Create decision points where you can assess progress and adjust course.

What's the cost of the learning? This is the traditional budget question, but framed differently. You're not projecting ROI; you're asking whether the learning is worth the investment required to acquire it. Sometimes expensive learning is worth it. Sometimes cheap experiments can teach you what you need to know.

What options does success create? If this initiative works, what becomes possible that wasn't possible before? What follow-on investments would make sense? An initiative that opens significant new possibilities has more strategic value than one that's a dead end even if it succeeds.

What's our exposure if it fails? How much are we risking? Is the downside bounded? Can we fail gracefully, or will failure be visible and damaging? The best AI investments have asymmetric payoffs: bounded downside, significant upside if things work.

Who will learn from this? Learning that stays with one team or one consultant has less value than learning that builds organizational capability broadly. How will insights from this initiative spread? Who will be better equipped afterward to lead the next initiative?

This framework won't give you precise numbers or guaranteed outcomes. But it will help you make better decisions under uncertainty—and it will help you explain those decisions to stakeholders who are understandably nervous about investing in a rapidly changing landscape

Investment Mistakes

Conversely, here are some patterns to avoid—mistakes I've seen repeatedly that undermine even well-intentioned AI programs.

Don't bet everything on one initiative. The temptation is to pick the highest-impact opportunity and focus all resources there. But in uncertain environments, concentration is dangerous. If your one big bet fails—and many AI initiatives do fail, especially early ones—you've lost your resources and your organizational confidence. A portfolio of smaller bets lets you learn across multiple fronts and limits the damage when individual experiments don't work.

Don't skip the unglamorous investments. The exciting AI applications get attention, but they often depend on unglamorous foundations: clean data, integrated systems, trained people, clear processes. Organizations that underinvest in foundations find their ambitious AI initiatives stalling on basic infrastructure problems. Budget for the boring stuff.

Don't ignore the change management costs. AI implementation isn't just a technology investment; it's an organizational change investment. People need to learn new tools, adapt workflows, sometimes accept that their roles are changing. These transitions require resources—training, communication, support, time. Organizations that budget only for the technology consistently underestimate the true cost of implementation.

Don't wait for perfect information. There will never be a moment when AI investment becomes obviously, risklessly correct. If you're waiting for certainty, you're waiting forever. The organizations that build AI capability are the ones that act under uncertainty, learn from the experience, and iterate. Analysis paralysis is its own form of failure.

Making the Investment Case

Even if you're convinced that AI investment is necessary, you may need to convince others—your colleagues, your board, your CEO, your executive team, your investors. And they've heard a lot of AI hype. They've seen initiatives that promised transformation and delivered disappointment. They're appropriately skeptical.

Communicating effectively about AI investment requires understanding what your audience is worried about—and addressing those concerns directly rather than talking past them.

Here are key questions you should be prepared to answer:

"Is this just hype?" Seasoned professionals have lived through technology hype cycles before. They've seen promising technologies fail to deliver, seen vendors overpromise, seen initiatives launched with fanfare and quietly abandoned. They're pattern-matching AI to previous disappointments. Your job is to acknowledge this concern directly. Yes, there's hype, and here's how we're cutting through it to focus on what really creates value.

"What if we invest and the technology changes?" The pace of AI development is genuinely disorienting. Capabilities that seemed years away arrive in months. Today's leading-edge platform becomes tomorrow's legacy system. Prudent leaders worry about investing in something that will be obsolete before it delivers value. The response: we're investing in learning and capability, not just technology. The specific tools will change; the organizational ability to use them effectively compounds over time.

"How do we know this won't be a money pit?" AI initiatives can consume resources without clear endpoints. Executives have seen IT projects that kept requiring "just a little more" investment without ever delivering promised returns. The response: we're structuring this as bounded experiments with clear decision points. We'll know by this date whether to scale, pivot, or stop. We're buying learning, not committing to an unlimited program.

"What about the risks?" Thoughtful leaders worry about bias, about privacy, about security, about reputational damage if AI goes wrong. A strong governance framework is your answer here. You're not asking them to accept unmanaged risk; you're showing them you've thought carefully about what could go wrong and how you'll prevent it.

"Can we afford this right now?" Budgets are always tight. There's always something else competing for resources. AI investment feels discretionary in a way that keeping the lights on doesn't. The response: reframe from "can we afford to invest?" to "can we afford not to invest?" Make the cost of inaction visible. Show what competitors are doing. Quantify the catch-up cost if you wait.

The most effective AI investment pitches don't oversell. They acknowledge uncertainty while making a compelling case for action despite that uncertainty. They address concerns directly rather than dismissing them. They frame AI as strategic necessity—as essential to future competitiveness—rather than as speculative bet on unproven technology.

And critically, they survive budget pressure. When economic conditions tighten and discretionary spending gets cut, AI investments framed as "innovation experiments" are vulnerable. AI investments framed as "building capabilities essential to compete" are more defensible. The framing you choose now determines whether your AI program survives the next downturn.

ASK AI

"Act as a skeptical investor. What concerns do you have about this proposal? What would you need to see to be convinced this is a wise use of resources?" Use the responses to strengthen your investment case.

The Investment Imperative

Let's close with a direct statement: AI investment is not optional for organizations that intend to remain competitive.

This is a "yes, and" moment. Yes, there's uncertainty—and here's how we act anyway. Yes, some investments will fail—and here's how we structure them to learn from failure. Yes, the technology will keep changing—and here's how we build organizational capability that compounds regardless of which specific tools win.

The executives who thrive in AI investment aren't the ones who predict outcomes accurately. They're the ones who structure investments to learn quickly, iterate effectively, and scale what works—while limiting downside when experiments don't pan out.

9

MAKING IT HAPPEN

Strategy is easy. Execution is hard.

You can have brilliant insights about AI's potential, a clear-eyed view of competitive dynamics, a compelling investment thesis, and a robust governance framework—and still fail completely if you can't make things happen in your organization. The graveyard of corporate initiatives is filled with strategies that were analytically sound yet failed to thrive.

This is the challenge: how to move from strategic intent to organizational execution. We'll cover how to assess where you're starting from, how to sequence your early actions, how to structure initiatives for learning, and how to navigate the inevitable resistance and setbacks. Most AI initiatives fail not because the technology doesn't work, but because the organization doesn't change. Understanding why—and what to do about it—is the difference between transformation that happens and transformation that's merely announced.

Let's start with an honest assessment of where you are.

Assessing Readiness

Before you can plan a journey, you need to know your starting point. Organizational readiness for AI spans multiple dimensions, and most organizations are stronger in some areas than others.

Technical readiness is what most assessments focus on: Do you have the data infrastructure? The technical talent? The systems integration capability? These matter, but they're rarely the binding constraint. Technical gaps can be filled with hiring, training, or partnerships. They're visible, tractable problems.

Process readiness asks whether your workflows can incorporate AI. Are decisions made in ways that could be augmented by AI input? Is work structured so that AI-generated outputs can flow into human processes? Or are your processes so rigid, so dependent on legacy systems and manual handoffs, that AI would need to work around them rather than within them?

Cultural readiness is the dimension most often underestimated and most likely to determine success or failure. Is your organization curious about AI or threatened by it? Do people have permission to experiment, or does failure carry career risk? Is there appetite for change, or is the organization exhausted from previous transformation initiatives? Do leaders model AI adoption, or do they delegate it to others while remaining personally disengaged?

Leadership readiness assesses whether the people who need to champion AI transformation are prepared to do so. Do your executives understand AI well enough to make good decisions? Are they willing to invest political capital in pushing through resistance? Will they stay the course when early results disappoint, or will they move on to the next priority?

TRY THIS

Assess your organization's AI readiness across all four dimensions: technical, process, cultural, and leadership. For each one, ask: What's our honest current state? What would need to be true for AI initiatives to succeed? What's the gap? Be ruthless about distinguishing between the organization you have and the organization you wish you had.

An honest readiness assessment often reveals uncomfortable truths. The organization that considers itself "data-driven" may discover its data is siloed, inconsistent, and poorly governed. The leadership team that talks enthusiastically about innovation may discover they've created a culture where nobody takes risks. The company that prides itself on technical excellence may discover its processes are calcified around legacy approaches.

These discoveries are valuable. You can't navigate from where you wish you were; you can only navigate from where you truly are. And the gaps you identify become the early work: not just implementing AI, but building the organizational capacity to implement AI effectively.

Preparing the Ground

The early period of any AI transformation sets patterns that persist. Move too slowly, and skeptics conclude you're not serious. Move too fast, and you outrun your organization's ability to absorb change. The first phase should accomplish several things simultaneously.

Build your coalition. Transformation doesn't happen through hierarchy alone. You need allies: executives who will advocate for resources, middle managers who will clear obstacles for their teams, technical leaders who will solve implementation problems, and influential individual contributors whose adoption will signal to others that AI is worth taking seriously. Identify these people early. Invest in their understanding and buy-in. Give them reasons to become champions rather than skeptics.

Establish credibility through action. Early in a transformation, people are watching to see if this is real or just another corporate initiative that will fade. Talking about AI's potential doesn't build credibility; demonstrating it does. Find ways to show, not tell — we'll discuss how in the next section.

Create safe spaces for experimentation. Most AI initiatives require people to try new things, which means risking failure. If failure carries career consequences, people won't try. Early in the transformation, you need to explicitly create permission to experiment—and that permission needs to be credible, not just stated. When early experiments don't work, how leadership responds will determine whether people keep trying or retreat to safety.

Surface and address resistance. Some people will resist AI adoption. Some resistance is based on legitimate concerns that deserve engagement. Some is based on fear or misunderstanding that can be addressed with information and support. Some is based on interests that conflict with the transformation's goals. You need to understand which is which. Resistance that's ignored doesn't disappear; it goes underground and sabotages implementation in ways that are harder to address.

Set realistic expectations. The worst thing you can do early is overpromise. If you set expectations for rapid transformation and then deliver slow, difficult progress, you lose credibility and create cynics. If you set expectations for a challenging journey that will require persistence, and then deliver exactly that, you maintain trust. The early period is when expectations are formed. Set them honestly.

Notice what's not on this list: deploying enterprise-wide AI systems, achieving transformative ROI, or solving your biggest business problems. Those come later. The first phase is about building an organizational foundation that makes everything else possible. I say this from experience — as an engineer, my initial instinct is to design a technical solution. I've learned the hard way that the best system deployed into unprepared soil doesn't take root

The Two-Tailed Problem

Most change management advice focuses on resistance—how to bring skeptics along, how to address fear, how to build momentum with reluctant adopters. That's necessary. But AI creates a different challenge: you're managing both tails of the distribution simultaneously.

On one end, you have people who want to halt AI entirely. They're worried about job loss, skeptical of the technology, protective of existing processes, or philosophically opposed. They need persuasion, evidence, and patience.

On the other end, you have people who want to move faster than the organization can safely absorb. They're deploying AI tools without governance review, making promises to customers before capabilities are proven, creating technical debt, or exposing the organization to risks they haven't fully considered. Their enthusiasm is genuine—but it's a different kind of problem.

This isn't a normal adoption curve where you're simply pulling the middle forward. The voices at both extremes are unusually loud, and they're often talking past each other. The enthusiasts dismiss the skeptics as Luddites; the skeptics dismiss the enthusiasts as reckless. Meanwhile, the pragmatic middle—people who could be productive adopters with the right support—feels caught between warring camps.

The polarization itself creates organizational dysfunction. When enthusiasts and skeptics are in open conflict, productive conversation becomes impossible. Each side caricatures the other. Middle-ground positions feel like betrayal to both camps.

Leaders need to actively bridge this divide:

Validate both concerns. The skeptics' concerns about risk, job impact, and unproven technology are legitimate. So is the enthusiasts' urgency about competitive pressure and opportunity cost. Acknowledge both rather than dismissing either.

Create shared experiences. Skeptics who participate in well-run pilots often become converts. Enthusiasts who see the complexity of enterprise deployment often develop more patience. Shared experience builds common ground that debate cannot.

Model the middle path. Demonstrate that it's possible to move with urgency and responsibility. Leaders who embody this balance give permission to others to occupy the same space.

Remember that this same polarization exists in your market. Some customers actively want AI-powered experiences—they're early adopters who seek out the latest capabilities. Others actively resist—they want human interaction, distrust AI, or have had bad experiences with poorly implemented automation. For some products and segments, AI is a selling point. For others, human service is the selling point. Know which is which, and don't assume the whole market feels the same way you do.

Follow the Pain

The best AI opportunities rarely announce themselves as AI opportunities. They show up as pain points, workarounds, and problems people have stopped complaining about because they've given up expecting solutions.

This is latent demand—needs that exist but aren't being expressed as requests for AI. Your employees aren't asking for "AI-powered document analysis." They're spending three hours every week manually cross-referencing contracts. Your customers aren't requesting "intelligent personalization." They're frustrated that they can't find what they need on your website. The demand is there; it's just wearing different clothes.

The opportunity is recognizing latent demand—seeing the friction, the jerry-rigged workarounds, the tasks people dread—and connecting it to capabilities that now exist.

Look for these signals:

The eternal backlog. Work that everyone agrees is valuable but never rises high enough in priority to get done. AI might make it tractable.

The heroic workaround. Processes that only work because someone stays late, maintains a secret spreadsheet, or performs manual gymnastics. That effort is a signal that something wants to be automated.

The abandoned request. Things customers or employees used to ask for but stopped requesting because the answer was always "we can't do that." Maybe now you can.

The "it would be nice" wish. Ideas that get dismissed as impractical or too expensive. Revisit them with fresh eyes—AI has changed what's possible.

This reframe also helps with positioning. "We're deploying AI" reinforces the two-tailed problem. "We're finally solving the contract review bottleneck" gets buy-in. The initiative is the same; the framing changes everything.

The richest opportunities aren't where people are demanding AI. They're where employees and customers have pain they've learned to live with.

Quick Wins and Strategic Bets

AI initiatives exist on a spectrum from quick wins to strategic bets. You need both, but they serve different purposes and require different approaches.

Quick wins are initiatives that can show results fast—weeks or months, not years. They may not be your highest-impact opportunities, but they're achievable with current capabilities and limited risk. Their primary value isn't the direct business impact, though that's nice. Their primary value is organizational: they demonstrate that AI works, they build confidence and capability, they create converts who experienced success and now want more, and they give you credibility to pursue larger initiatives.

Choose quick wins deliberately. The best ones are visible (people will see the results), demonstrable (you can show before-and-after), and relevant (they connect to work people care about). An AI that helps the sales team prepare for meetings faster is a better quick win than an AI that optimizes backend data processing, even if the latter has higher theoretical value. People need to see AI making their lives better, not just hear about efficiency gains in areas they don't touch.

Strategic bets are larger initiatives aimed at significant business impact. They take longer, cost more, and carry more risk. They're also where the real transformation happens. Quick wins keep the lights on politically and build organizational capability; strategic bets change the trajectory of the business.

The relationship between quick wins and strategic bets is sequential and mutually reinforcing. Quick wins build the credibility, capability, and organizational appetite that make strategic bets possible. Strategic bets, when they succeed, validate the overall direction and justify continued investment. An organization that pursues only quick wins never transforms. An organization that pursues only strategic bets often fails before those bets pay off, because it lacks the organizational foundation to sustain long-term initiatives through inevitable setbacks.

Think of your AI initiatives like a venture capital portfolio: many small experiments, some medium-sized initiatives, a few big bets. Most of the small experiments will fail to produce breakthroughs, but they're cheap and they generate learning. The medium initiatives have better odds and higher stakes. The big bets are where transformation lives—but you only take them when you've built the capability and conviction to execute.

Build, Buy, or Partner?

You face an important question when resourcing your AI initiatives: Do you build the capability internally, buy it from vendors, or partner with external organizations? The answer depends on several factors and getting it wrong can be expensive.

Build internally when the capability is core to your competitive differentiation, when you need deep customization that vendors can't provide, when you have (or can develop) the talent to execute, and when you're willing to invest the time it takes. Building creates the deepest organizational learning and the most defensible capability—but it's slow, expensive, and requires talent that's in high demand everywhere.

Buy from vendors when the capability is well-understood and commoditized, when speed matters more than customization, when you lack the talent to build, or when the capability isn't core to your differentiation. Vendor solutions let you move fast and avoid reinventing wheels. The risks are dependency, limited customization, and capabilities that competitors can also buy.

Partner when you need capabilities you can't build but that aren't available as off-the-shelf products, when you want to share risk and investment, or when a partner brings complementary strengths. Partnerships can accelerate progress significantly—but they require trust, aligned incentives, and clear governance. Failed partnerships are expensive in time, money, and organizational energy.

A few principles help navigate this decision:

Don't build what you can buy, unless it's core to your differentiation. Engineering resources are scarce and expensive. Using them to build commodity capabilities is waste. Save your building capacity for things that differentiate you. Remember: "we could build this" is not the same as "we should build this."

Don't buy what you need to own. If a capability is genuinely strategic—if it's core to how you compete—depending on a vendor is dangerous. They control the roadmap, the pricing, and ultimately whether the capability continues to exist. Strategic capabilities need to be owned, even if ownership is expensive.

Don't outsource your learning. Regardless of whether you build, buy, or partner, make sure your organization is learning. If all the knowledge about how AI works in your context sits with vendors or partners, you're not building capability—you're renting it. Structure relationships so that internal learning happens even when external parties do the work.

Principles for Tool Selection

When you've decided to buy or partner rather than build, you face a crowded marketplace of AI tools—each promising transformation, most delivering something more modest. A few principles help cut through the noise.

Start with the problem, not the tool. What are you actually trying to accomplish? The best tool depends on your specific needs, your existing systems, your team's capabilities, and your risk tolerance. "We need an AI strategy" isn't a problem statement. "We're losing deals because our proposal turnaround is too slow" is. Start with the business problem; let that drive tool selection.

Experiment before committing. Most tools offer trials or limited free tiers. Use them. But test with real work, not demo scenarios. The slick presentation doesn't tell you how the tool performs on your actual data, with your actual users, in your actual workflows. Involve the people who will use the tool in the evaluation—not just the people buying it.

Consider the ecosystem. A tool that integrates well with your existing systems may be more valuable than a technically superior tool that operates in isolation. Integration friction is real and often underestimated. The best AI tool that doesn't connect to your CRM, your document management system, or your communication platforms creates islands of capability rather than transformed workflows.

Plan for change. Whatever tool you choose today will likely be superseded by something better. The AI landscape is evolving too fast for any choice to be permanent. Avoid deep lock-in where possible. Build organizational capability that transfers across tools—prompt fluency, judgment about AI outputs, integration patterns—not just expertise in a single platform. The skills matter more than the specific tool.

Remember the human side. The best tool is one people will actually use. Consider user experience, training requirements, and change management—not just technical capabilities. A sophisticated tool that your team resists or misuses is worse than a simpler tool they embrace. Adoption is a feature.

Be skeptical of "specialized" claims. The market is crowded with domain-specific AI tools—legal AI, healthcare AI, financial AI—that are essentially wrappers around general AI with some tailored prompting. That's not necessarily bad; packaging and workflow integration have value. But make sure you're paying for genuine additional capability, not just convenient presentation of something you could already have with your existing resources.

The goal is to make good-enough choices that let you start learning, while maintaining the flexibility to adapt as better options emerge. In a fast-moving landscape, the ability to course-correct matters more than getting it right the first time.

Whatever path you choose, the next question is how to test whether it's working. That's where pilot design comes in.

Designing Effective Pilots

Most organizations run pilots. Few design them well. A well-designed pilot answers the questions you need answered. A poorly designed pilot generates activity without insight—and often leads to wrong conclusions.

Common pilot design failures:

The rigged pilot. The team running the pilot is so invested in success that they choose the easiest conditions, provide extra support, and declare victory based on results that won't generalize. The pilot "succeeds," but scaling fails because real conditions are harder than pilot conditions.

The orphan pilot. The pilot runs in isolation, disconnected from the people and processes it would need to integrate with at scale. It proves the technology works in a vacuum but reveals nothing about whether it works in your actual organization.

The indefinite pilot. The pilot keeps running without clear success criteria or decision points. It becomes a permanent small project—never killed, never scaled, never producing the learning it was supposed to generate. "We're still piloting" becomes an excuse for not making decisions.

The unmeasured pilot. The pilot runs without baseline metrics or clear measurement of outcomes. When it ends, no one can say definitively whether it worked. Decisions about scaling are based on impressions and anecdotes rather than evidence.

Well-designed pilots share certain characteristics.

They have clear hypotheses. Not "let's see if AI can help with customer service," but "we believe AI can reduce average handle time by 20% while maintaining satisfaction scores." The hypothesis might be wrong—that's fine. But without a hypothesis, you can't design a test that produces useful learning.

They have realistic conditions. The pilot should run in conditions that resemble scale deployment. Real users, real data, real processes. If the pilot requires special conditions to succeed, you're not learning whether the initiative will work—you're

113

learning whether it can work in a carefully controlled environment, which is a different question.

They have clear decision criteria. Before the pilot starts, define what success looks like and what decisions you'll make based on outcomes. If metrics exceed X, we scale. If they fall below Y, we stop. If they're between X and Y, we iterate with specific changes. This prevents post-hoc rationalization and indefinite piloting.

They have bounded timelines. A pilot should have a defined end date. Open-ended pilots become permanent projects. The timeline creates urgency and forces decisions.

They build organizational capability, not just technical validation. A pilot that proves the technology works but leaves no one in the organization able to operate it has limited value. Design pilots so that internal people are deeply involved—learning, building skills, understanding what works and what doesn't. The organizational learning is as valuable as the technical validation.

TRY THIS

Identify your AI adopters, skeptics, and blockers. For each group, ask: What would it take to move them one step forward? Your early adopters need resources and visibility. Your skeptics need evidence and safe ways to experiment. Your blockers need either to be convinced or routed around. Knowing who's who is the first step to building a coalition that can drive change.

Failure Modes

Understanding how AI initiatives fail helps you avoid the common traps. Here are patterns I've seen repeatedly—each with an organizational dimension that's usually the real culprit.

The Pilot That Never Scales. A successful pilot is declared, but somehow it never makes the transition to broader deployment. There's always a reason to delay: the technology needs refinement, the organization isn't ready, there are other priorities. Time passes. The pilot team eventually disperses. The organization learns nothing except that AI initiatives don't go anywhere. *The organizational reality:* Often there was never real commitment to scale. The pilot was approved because it was low-risk, but scaling requires political capital, budget, and change that leadership wasn't prepared to drive. The pilot succeeded, but the organizational will to act on it didn't exist.

Death by Committee. Every AI initiative requires approval from multiple stakeholders, each with veto power and different concerns. Legal worries about liability. IT worries about security. Operations worries about disruption. HR worries about workforce impact. No single stakeholder can approve, but any can block. Initiatives die in endless review cycles or emerge so compromised they can't succeed. *The organizational reality:* The organization has optimized for risk avoidance rather than value creation. Governance has become obstruction. Without executive intervention to break logjams and accept calculated risks, nothing significant can happen.

The Technology Solution Looking for a Problem. Someone becomes enamored with an AI capability and goes looking for ways to apply it, rather than starting with business problems and asking how AI might help. The resulting initiatives are technically interesting but disconnected from what the business needs. They struggle to demonstrate value because value wasn't the starting point. *The organizational reality:* Technical enthusiasm has outrun business discipline. The organization lacks processes for connecting AI capabilities to business priorities. Technology and business teams aren't collaborating effectively.

Innovation Theater. The organization announces AI initiatives with great fanfare. There are press releases, internal communications, maybe a dedicated "AI lab" or "innovation center." But the initiatives are disconnected from core business operations. They're showcases, not transformations. The real work of the organization continues unchanged. *The organizational reality:* AI has been delegated to a sideshow while the main business remains protected from disruption. Leadership gets credit for innovation without accepting the difficulty of actual change. Eventually the theater closes and everyone pretends it never happened.

A note on innovation theater: AI labs and innovation centers aren't inherently theater. Sometimes they're exactly what's needed to foster creative development. They become theater when they're disconnected from core operations — when they exist to signal innovation rather than drive it. The test is simple: does work from your innovation center change how the rest of the organization operates? If so, it's infrastructure. If not, it's a stage set.

Change Fatigue. The AI initiative arrives in an organization exhausted from previous transformation efforts. People have seen initiatives come and go. They've learned that the safest strategy is to wait things out—keep your head down, do the minimum required, and eventually leadership will move on to something else. The initiative never gains traction because the organization has developed antibodies against change. *The organizational reality:* Previous transformations may have been poorly managed, over-promised, or abandoned mid-stream. Trust has been damaged. Rebuilding it requires acknowledging the history, demonstrating that this time is different, and proving commitment through sustained action rather than just announcements.

Each of these failures has a technical surface and an organizational core and makes change management essential.

The J-Curve

Here's something most AI vendors won't tell you: performance usually gets worse before it gets better.

When you introduce AI into existing workflows, there's a learning curve. People need to figure out how to use new tools. Processes need to be adjusted. Integration issues need to be resolved. During this period, productivity often drops. The old way no longer works smoothly because it's being disrupted; the new way doesn't work smoothly yet because it's not fully implemented. You're in the valley.

This is the J-curve of AI adoption. Performance dips, then recovers, then exceeds the original baseline. The curve looks like a J tipped on its side—down first, then up. Understanding this pattern is crucial because the valley is where most initiatives die.

The J-Curve

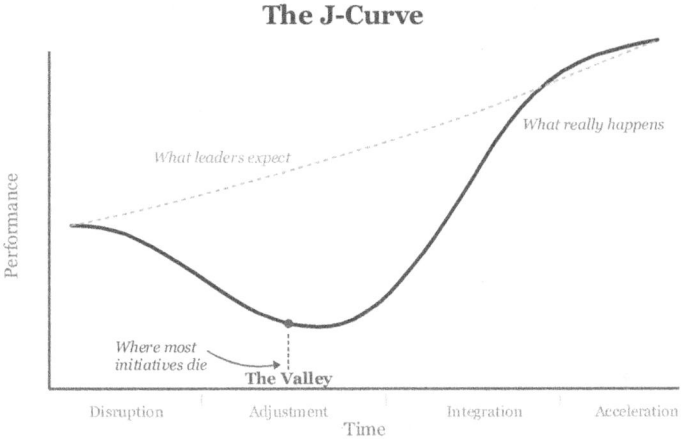

Here's what the J-curve looks like in practice. A sales team implementing AI-assisted prospecting might see their close rates drop for the first six weeks as reps learn new workflows, figure out which AI suggestions to trust, and adjust their processes. By month two, they're back to baseline. By month three, they're exceeding their previous performance by 15%. But if leadership had pulled the plug at week four—when the numbers looked worst—they would have concluded that "AI doesn't work for our sales team" and abandoned an initiative that was succeeding.

Organizations that don't expect the valley panic when they hit it. They see productivity dropping, hear complaints from frustrated users, watch early metrics disappoint—and they conclude the initiative is failing. They pull back support, cut resources, or kill the project entirely. They never reach the upward part of the curve because they bailed out in the valley.

Organizations that expect the valley plan for it. They set expectations with stakeholders that performance will dip before it improves. They budget extra time and resources for the transition

117

period. They measure progress in leading indicators (adoption, learning, capability building) rather than just lagging indicators (productivity, efficiency). They celebrate milestones that indicate they're climbing out of the valley, not just ultimate outcomes.

The J-curve also explains why quick wins matter so much. Quick wins that show AI working—even in limited contexts—provide evidence that the upward slope is real. They give people confidence to persist through the valley in larger initiatives. They counter the narrative that "AI doesn't work here" that can take hold when all anyone sees is the downward slope.

Practical implications for navigating the J-curve:

Prepare people in advance. Tell them the valley is coming. When they hit it, they should recognize it as an expected phase, not evidence of failure. "We're in the valley" is a different message than "this isn't working."

Permission to fail. AI efforts may be clumsy. You'll underestimate capabilities in some areas and overestimate them in others. Your team will make mistakes, which is good because that's how learning works. You can't learn from mistakes you never make. This is permission to be bad at AI. Not forever. Not carelessly. But deliberately embracing the early awkwardness as the price of eventual fluency.

Provide extra support during the transition. The valley is when people need the most help—training, troubleshooting, encouragement. This is exactly when you're tempted to cut resources because results are disappointing. Resist that temptation. Increased support during the valley shortens the valley.

Track leading indicators. Lagging indicators like productivity will look bad in the valley. Leading indicators like adoption rates, user proficiency, and declining support requests can show you're making progress even when output metrics are still suffering.

Celebrate valley milestones. Recognize progress through the difficult period. "We've gotten past the initial learning curve" is worth celebrating, even if ultimate productivity gains haven't materialized yet.

Don't declare victory(or defeat) too early. The valley creates pressure to make definitive judgments. Resist. An initiative that's in the valley isn't failing; it's transitioning. Give it time to reach the upward slope before drawing conclusions.

TRY THIS

Map your current AI initiatives against the J-curve. Which are in the valley right now? What support do they need to get through it? What leading indicators could show progress even when lagging indicators look discouraging? Have you prepared stakeholders to expect the valley, or are they likely to interpret it as failure?

The J-curve is why sustained leadership commitment matters so much. Executives who champion an AI initiative need to stay committed through the valley—defending resources, maintaining expectations, and providing air cover for teams doing difficult work. If leadership abandons initiatives at the first sign of trouble, the organization learns that AI projects are risky to one's career, and future initiatives become harder to staff with talented people willing to take that risk.

Keep Moving

We've covered a lot of ground, so let's recap what matters most.

AI transformation is organizational transformation. The technology is the easy part. The hard part is changing how people work, what they believe is possible, and what your organization is capable of. If you treat AI implementation as a technical project, you'll fail. If you treat it as an organizational change effort that happens to involve technology, you have a chance.

Start where you are. Honest assessment of readiness—including the uncomfortable parts—is the foundation for realistic planning. Move fast enough to matter, slow enough to bring people along. And expect the valley: performance will dip before it improves, and the organizations that transform are the ones that persist through difficulty.

The best AI strategy is the one you execute. Start. Learn. Iterate. Scale what works. Kill what doesn't. Keep moving.

Part III has offered frameworks for leading AI responsibly: governance to manage risk, investment approaches for uncertain returns, and a playbook for execution. What remains is to look forward—to consider what's coming, what's uncertain, and how to build an organization that can navigate whatever the future holds.

That's where we conclude.

Conclusion

THE ROAD AHEAD

We began this book in a Waymo, watching science fiction become Tuesday. We've covered a lot of ground since then: shifting from scarcity to abundance, what AI actually is and isn't, building fluency, reshaping competition and business models, the changing nature of work, governance and ethics, investment under uncertainty, and an organizational playbook for making transformation real.

Now let's look ahead. What's coming? What's uncertain? And how do you build an organization—and a leadership practice—that can navigate whatever emerges?

What We Can See

Some things about the AI future are reasonably predictable—not because anyone has a crystal ball, but because the trajectories are already visible.

AI capabilities will continue to expand. The systems available today will look primitive compared to what's available in three years, just as the systems from three years ago look primitive today. This isn't speculation; it's extrapolation from a clear trend. The specific capabilities that emerge may surprise us, but the fact of continued advancement won't.

Costs will continue to fall. The economics we previously discussed—plummeting inference costs, increasingly capable open-source models, commoditization of what was recently cutting-edge—will accelerate. Capabilities that feel expensive today will feel cheap tomorrow. This democratization will put AI within reach of organizations that currently feel priced out.

121

Integration will deepen. AI will move from being something you occasionally consult to something embedded in how work gets done. The distinction between "using AI" and "just working" will blur. For knowledge workers, AI assistance will become as unremarkable as spell-check—present everywhere, noticed nowhere. In the physical world, the change may take longer, but it's coming as well.

The talent landscape will shift. Skills that command premiums today may be commoditized tomorrow. New skills will emerge that we can't fully anticipate. The organizations that invest in continuous learning—that treat AI fluency as a journey rather than a destination—will adapt more gracefully than those that don't.

Regulation will evolve. Organizations that build governance capacity now will be better positioned as requirements shift than those scrambling to catch up.

These trends are visible enough that planning for them is reasonable. The organizations that act on them—that invest in capability building, that develop governance frameworks, that cultivate adaptive cultures—will have advantages over those that wait for certainty.

What We Can't See

But much remains genuinely uncertain—and intellectual honesty requires acknowledging what we don't and can't know.

We don't know how far current approaches will scale. The AI advances of recent years have come largely from scaling up existing architectures with more data and more compute. Whether this scaling continues to yield dramatic improvements, or whether we hit walls that require fundamentally new approaches, is an open question. Experts disagree. The answer matters enormously for what AI can ultimately do, but no one knows for certain.

We don't know which industries will be most disrupted, or when. It's easy to generate plausible scenarios where AI transforms almost any sector. It's harder to know which transformations will happen in two years versus twenty, which will be gradual versus sudden, which predictions will prove prescient versus embarrassingly wrong. Humility is warranted.

We don't know what new capabilities will emerge, or what they'll make possible. The history of technology is full of capabilities that seemed impossible until they weren't, and applications that no one anticipated until they existed. AI will likely follow this pattern. Some of the most important developments will be ones we can't currently imagine.

We don't know how society will adapt. Technology doesn't determine social outcomes; it creates possibilities that humans then navigate through politics, culture, and collective choice. How we handle AI's impact on employment, on inequality, on privacy, on the nature of truth itself—these are not technical questions with technical answers. They're human questions that will be answered through human processes, for better or worse.

This uncertainty isn't a reason for paralysis. It's a reason for building adaptive capacity—the ability to respond effectively to developments you didn't predict. The goal isn't to guess the future correctly. It's to build an organization that can thrive across a range of possible futures.

The Bubble and the Revolution

If you're reading this during an AI market correction—or even after one—resist the temptation to declare the revolution over.

Every transformational technology in modern history has followed a similar arc: explosive growth, speculative excess, painful correction, continued transformation. The pattern is so consistent it's almost a law.

Railroads revolutionized transportation and commerce in the nineteenth century—and the railroad bubble of the 1840s wiped out investors and collapsed companies across Britain and America. The technology kept transforming the world anyway. Electricity changed everything about how we live and work—and the utility bubble of the early twentieth century destroyed fortunes. The lights stayed on.

The dot-com crash of 2000-2001 vaporized trillions of dollars in value. Yet, the biggest impact of the internet was still ahead. In 2001, most people were still on dial-up. The iPhone was six years away. Social media didn't exist. Cloud computing was mostly theoretical. Streaming was a niche. And almost nobody saw it because the hype had moved on.

AI will likely follow this pattern. Market corrections are about prices, not about underlying technological reality. Crashes make the surviving companies more disciplined. It clears out the speculation and leaves behind the substance. The transformation continues.

The executives who panic during corrections—who cut AI investments, abandon initiatives, and retreat to the familiar—will lose ground to those who stay the course. Not because staying the course is always right, but because abandoning strategic direction based on market sentiment is almost always wrong. The signal is the technology. The noise is the market. Don't confuse them.

ASK AI

"If AI valuations crashed 50% tomorrow, how would you defend continued AI investment to a skeptical board?" Then practice your answer. This isn't just an exercise—it's preparation for a conversation you may need to have.

Yes, And—Through Everything

We started this book with "yes, and"—the improv principle of accepting reality and building on it. I want to return to it now, because it's not just a mindset for understanding AI. It's a mindset for leading through whatever comes next.

Yes, the technology will keep changing—and here's how we build learning capacity that compounds regardless of which specific tools win.

Yes, some investments will fail—and here's how we structure them to learn from failure rather than be destroyed by it.

Yes, markets may correct—and here's how we maintain strategic direction through the noise.

Yes, we face genuine uncertainty—and here's how we build organizations that can adapt to futures we can't predict.

Yes, this is hard—and here's how we persist through the valley, knowing the upward slope is real.

The alternative is blocking. "AI is overhyped." "The correction proves it was never real." "Let's wait until things settle down." Each of these statements might feel true in the moment. Each is a way of refusing to engage with reality as it truly is.

Leaders who block will watch from the sidelines as others build. They'll feel vindicated during corrections and confused when the transformation continues anyway. They'll eventually be forced to engage—but later, from behind, with less capability and less time.

Leaders who practice "yes, and" will make mistakes. They'll invest in things that don't pan out. They'll sometimes move faster than their organizations can absorb. But they'll be learning, adapting, building. When the future arrives—whatever future it turns out to be—they'll be ready to meet it.

Revisiting the List

In the Introduction, you wrote down three things you believed AI couldn't do well.

Look at it now. How has your view changed?

Perhaps you've discovered that AI can do some of those things better than you expected—not perfectly, not without oversight, but better than you would have guessed before you started this journey.

Perhaps you've refined your understanding of what "doing it well" means, and found that the line between human and AI capability is blurrier than you thought.

Or perhaps you've confirmed that some things really do remain beyond AI's current reach—and you now have a clearer sense of why, and what that means for where human value lies.

Either way, you know more than when you started. You have frameworks for thinking about AI strategically. You understand the shift from scarcity to abundance, and what it means for value propositions. You know how to evaluate AI capabilities and limitations. You have a path for building your own fluency, and tools for leading your organization's transformation.

That list of "things AI can't do" will keep evolving. Capabilities that seem impossible today will become routine. New limitations will emerge that we can't currently anticipate. The list is never final— and that's exactly the point.

Leading AI well isn't about having all the answers. It's about having the frameworks to engage productively with questions that keep changing.

Ask AI

"What AI capabilities should I expect to be commonplace in my industry in three years?" Compare the answer to what seemed impossible three years ago to help calibrate your intuitions for the pace of change.

Block or Build?

We began the book with a question: will you block or build?

The future isn't something that happens to us. It's something we help create. Every decision we make about AI—every investment, every governance choice, every organizational change, every personal commitment to building fluency—shapes the future we will inhabit.

Market corrections don't end revolutions. Blocking does.

So don't block. Build.

And keep saying "yes, and."